Designing Embedded Systems with Arduino

Tianhong Pan · Yi Zhu

Designing Embedded Systems with Arduino

A Fundamental Technology for Makers

Springer

Tianhong Pan
Jiangsu University
Zhenjiang
China

Yi Zhu
Jiangsu University
Zhenjiang
China

ISBN 978-981-10-4417-5 ISBN 978-981-10-4418-2 (eBook)
DOI 10.1007/978-981-10-4418-2

Library of Congress Control Number: 2017937915

Printed on acid-free paper

This Springer imprint is published by Springer Nature
The registered company is Springer Nature Singapore Pte Ltd.
The registered company address is: 152 Beach Road, #21-01/04 Gateway East, Singapore 189721, Singapore

Preface

Our world is full of smart and connected products embedded with processors, sensors, and software. Do-it-yourself communities have always been fascinated by the fact that a person can design and build his/her own smart system for specific tasks. Arduino presents us with an alternative platform to build such amazing products. Users can download the Arduino Integrated Development Environment (IDE) and code their own program using the C/C++ language as well as the Arduino Core library, which provides a considerable amount of helpful functions and features. Arduino makes it easy to sense and respond to touch, sound, position, heat, light, etc.

The SPIED (Summer Program for Innovative Engineering Design) has been implemented in three countries, i.e., Japan, China, and Korea, on a rotation basis since 2013. The role of SPIED is to establish innovative engineering education in the three countries. In the SPIED, senior-level and graduate students from Japan, China, and Korea stay and work together on planning, designing, production, and presentation of a prototype mechatronics and robotic system. By combining engineering design technique with the ability to identify problems from a multidisciplinary perspective, SPIED provides participants with a sense of achievement when they undergo the process of drawing their dreams as a concept, followed by designing and creating them as prototypes. However, mechatronics and robotic systems involve numerous techniques related to multiple disciplines. Students need to spend a considerable amount of time learning technologies. A unique advantage of Arduino is that it can be used by anyone, even people with no programming or electronics experience. Arduino is an open-source platform composed of very simple and easy-to-use hardware and software, which has mainly been developed for prototyping purposes. Therefore, it is a great fit for students.

In this book, we want to systematically integrate Arduino modules with Arduino platform and train beginners in understanding this technology. Furthermore, information on various topics including sensors, photics, electronics, mechatronics, mathematical calculations, etc. is also introduced in this book, which can help readers explore system development from an interdisciplinary perspective.

Objective and Intended Audience

The purpose of this book is to present programming and electronics techniques based on Arduino and to discuss them from the point of view of using micro controller technology to interact with the environment. Over the last three years, notes based on this book have been used to support the Summer Program for Innovative Engineering Design (SPIED), which has been implemented by three countries, Japan, China, and Korea, on a rotation basis (http://ire-asia.org/ire/spied/). The book can also be used in senior-level/first-year-graduate courses on micro-controllers and its applications. Portions of these notes have been used to support training courses for electronics makers and hobbyist.

Book Contents

Although this is a book on open-source hardware and electronics, you will find a number of code examples. They are used to configure the hardware as desired and make it do what we want it to do. The authors are a professional teacher with a good experience in Embedding System Design. Through our partnership, we try to show a model of how traditional education can merge with the makers of the world to create a much richer learning experience than is possible to have by learning passively. Chapters 1–6 are written by Prof. Tianhong Pan, and Chaps. 7 and 8 are written by Biqi Sheng Ph.D. and Prof. Yi Zhu respectively.

The book begins (Chap. 1) by pointing out the different variants of Arduino boards. Next, Arduino history and characteristics are quickly reviewed, and the driver installation procedure and IDE of Arduino are also introduced.

Chapter 2 describes many embedded basic functions, such as the functions for reading and writing digital and analog input and output pins, interrupt functions, mathematic functions, and serial communication functions.

Chapter 3 presents the various types of sensor modules available for Arduino. It covers many of the commonly available types, such as the temperature sensor, joystick module, analogy sound sensor, and other items that are not specific to Arduino, but are compatible. Electrical pin-out information, schematics, and software are provided for many of the items discussed.

Chapter 4 explains how you can make things move by controlling motors with Arduino. A wide range of motor types are covered: DC motor, servo, stepper motor. All kinds of driving circuits and their schematics are introduced in this chapter.

Chapter 5 focuses on wireless techniques such as: infrared transmitter/receiver Bluetooth, ZigBee, Wi-Fi, etc. The examples in this chapter demonstrate how to connect Arduino to devices and modules and realize remote control.

Chapters 6–8 cover some projects that illustrate the capabilities of Arduino boards and sensor modules. They are intended to demonstrate how Arduino can be applied in various situations. Each example description includes theory of operation, schematics, detailed parts lists, layouts, and an overview of the software necessary for it to function.

Zhenjiang, China

Tianhong Pan
Yi Zhu

Acknowledgements

We would like to thank all the reviewers of the earlier drafts of this book, who made very useful comments and suggestions. In particular, we would like to thank Prof. **Zhongwei Jiang** from the Yamaguchi University, Japan, for his comments and encouragement. We also gratefully acknowledge many current students at Jiangsu University who have carefully provided reviews and performed tests: **Biqi Sheng, Wei Guo, Jiao Chen, Jiali Ding, Tao Jiang**, etc.

In the spirit of continuous improvement, we encourage feedback from students, faculty, and practitioners who use this book. We hope you find it useful.

Contents

Part II Case Studies

About the Authors

Tianhong Pan obtained his Ph.D. degree in April, 2007 from Shanghai Jiao Tong University, China. He then worked for 1.5 years at National Tsinghua University, Taiwan China since 2009, and 3.5 years at University of Alberta, Edmonton Canada since 2011 as a postdoctoral fellow, where gained valuable industrial experience in semiconductor manufacturing processes and petrochemical industry. Now, He has been a professor in School of Electrical and Information Engineering, Jiangsu University, Zhenjiang, China. As the core member of innovative research & education organization of Asia, he organized all Summer Program of Innovative Engineering Design since 2013 and got a good experience in "Embedded System Design using Arduino". In addition to teaching Arduino, he enjoys automating everyday tasks, designing and etching PCBs, and lots of random things in between. Much of his time has been spent researching, building, and testing various controllers to make his device work automatically. Additionally, he published several articles in refereed journals and international conferences.

Yi Zhu received the Ph.D. degree in communication and information system from Nanjing University of Posts and Telecommunications, Nanjing, China, in 2016. He is currently an Associate Professor of School of Computer Science and Communication Engineering, Jiangsu University, Zhenjiang, China. Since 2010, he has started working in innovative engineering education and maker training. As the core member of Innovative Research & Education Organization of Asia, he attended all Summer Program of Innovative Engineering Design and Creative Engineering Design Competition since 2013. In 2014 and 2015, he visited Yamaguchi University of Japan as visiting lecturer to teach "Embedded System". These abroad education activities provided great valuable experiences for forming his education idea of maker training. Through teaching practice, he proposed an effective maker training mode based on "Arduino/Raspberry Pi/Android/Robot" in the end of 2015. For his contribution in maker education, he received the first prize of Teaching Achievement Award by Jiangsu University and the first prize of National Engineering Training by Ministry of Education institution of Engineering Training Guidance Committee respectively in 2016.

Part I
Basic Skill Training and Application

Chapter 1
Getting Started with Arduino

1.1 Introduction

In 2005, Massimo Banzi, David Cuartielles, Tom Igoe, Gianluca Martino, and David Mellis came up with an idea for an easy-to-use programmable device for interactive art design projects at the Interaction Design Institute Ivrea in Ivrea, Italy. The device needed to be simple, easy to connect to various things (such as relays, motors, and sensors), and easy to program. It also needed to be inexpensive to make it cost-effective for students and artists. They selected an AVR family of 8-bit microcontroller (MCU or μC) devices from Atmel and designed a self-contained circuit board with easy-to-use connections, wrote bootloader firmware for the microcontroller, and integrated it all into a simple development environment that used programs called "sketches." The result was Arduino.

Arduino is an open-source microcontroller that enables programming and interaction; it is programmed in C/C++ with an Arduino library to allow it to access the hardware. This allows for more flexible programmability and the ability to use electronics that can interface with Arduino. Because Arduino is open source, the plans for the circuits are available online for free to anyone who wants to use and create their own board based on the schematics, as long as they share what they create. This allows for considerable customizability in projects; till date, users have built Arduinos of different sizes, shapes, and power levels to control their projects. Arduino is composed of two major parts:

1. The Arduino board, which is a piece of hardware you work on when you build your objects.
2. The Arduino IDE, which is a piece of software you run on your computer. You use the IDE to create a sketch (a small computer program) that you upload to the Arduino board.

Arduino is different from other platforms in the market because of the following features:

T. Pan and Y. Zhu, *Designing Embedded Systems with Arduino*,
DOI 10.1007/978-981-10-4418-2_1

1. It is a multiplatform environment; it can run on Windows, Macintosh, and Linux.
2. It is based on a processing programming IDE, which is an easy-to-use development environment used by artists and designers.
3. You program it via a USB cable, not a serial port. This feature is useful, because many modern computers do not have serial ports.
4. It is open-source hardware and software—if you wish, you can download the circuit diagram, buy all the components, and make your own Arduino board, without paying anything to the makers of Arduino.
5. The hardware is cheap.
6. There is an active community of users, so there are many people who can assist you.
7. The Arduino project was developed in an educational environment, and is therefore, great for newcomers to get things working quickly.

Owing to these special features, there are many potential applications:

1. Real-world monitoring
   - Automated weather station
   - Lightning detection
   - Sun tracking for solar panels
   - Background radiation monitor
   - Automatic wildlife detector
   - Home or business security system
2. Small-scale control
   - Small robots
   - Model rockets
   - Model aircrafts
   - Quadrotor UAVs
   - Simple CNCs for small machine tools
3. Small-scale Automation
   - Automated greenhouse
   - Automated aquarium
   - Laboratory sample shuttle
   - Precision thermal chamber
   - Automated electronic test system
4. Performance Art
   - Dynamic lighting control
   - Dynamic sound control
   - Kinematic structures
   - Audience responsive artwork

1.2 Arduino Variants

Arduino is rapidly becoming one of the most popular microcontrollers used in robotics. There are many different types of Arduino microcontrollers that differ not only in design and features, but also in size and processing capabilities. However, there are only two models that use completely different chips: the Standard and the Mega. The Standard is the basic Arduino that uses the Atmega8/168/328 chip, whereas the Mega is a different Arduino board with more I/O pins and uses the beefier Atmega1280 chip.

The makers of Arduino also developed software that is compatible with all Arduino microcontrollers. The software, also called "Arduino," can be used to program any of the Arduino microcontrollers by selecting them from a drop-down menu. Being open source, and based on C, Arduino users are not necessarily restricted to this software, and can use a variety of other software to program their microcontrollers.

There are many additional manufacturers who use open-source schematics provided by Arduino to make their own boards (either identical to the original, or with variations to add to the functionality), e.g., DFRobot.

In this chapter, the Arduino Uno R3 and DFRobot Romeo BLE boards are introduced.

1. **Arduino Uno R3**

The Arduino Uno is a microcontroller board based on the ATmega328. It has 14 digital input/output pins (of which 6 can be used as PWM outputs), 6 analog inputs, a 16 MHz ceramic resonator, a USB connection, a power jack, an ICSP header, and a reset button. It contains everything needed to support the microcontroller; simply connect it to a computer with a USB cable or power it with an AC-to-DC adapter or battery to get started. Central to the Arduino interface board, shown in Fig. 1.1, is an onboard microcontroller.

Specifications of the Arduino UNO R3 are as follows:

- Microcontroller: ATmega328
- Operating Voltage: 5 V
- Input Voltage (recommended): 7–12 V
- Input Voltage (limits): 6–20 V
- Digital I/O Pins: 14 (of which 6 provide PWM outputs)
- Analog Input Pins: 6
- DC Current per I/O Pin: 40 mA
- DC Current for 3.3 V Pin: 50 mA
- Flash Memory: 32 KB of which 0.5 KB is used by the bootloader
- SRAM: 2 KB (ATmega328)
- EEPROM: 1 KB (ATmega328)
- Clock Speed: 16 MHz (Fig. 1.2)

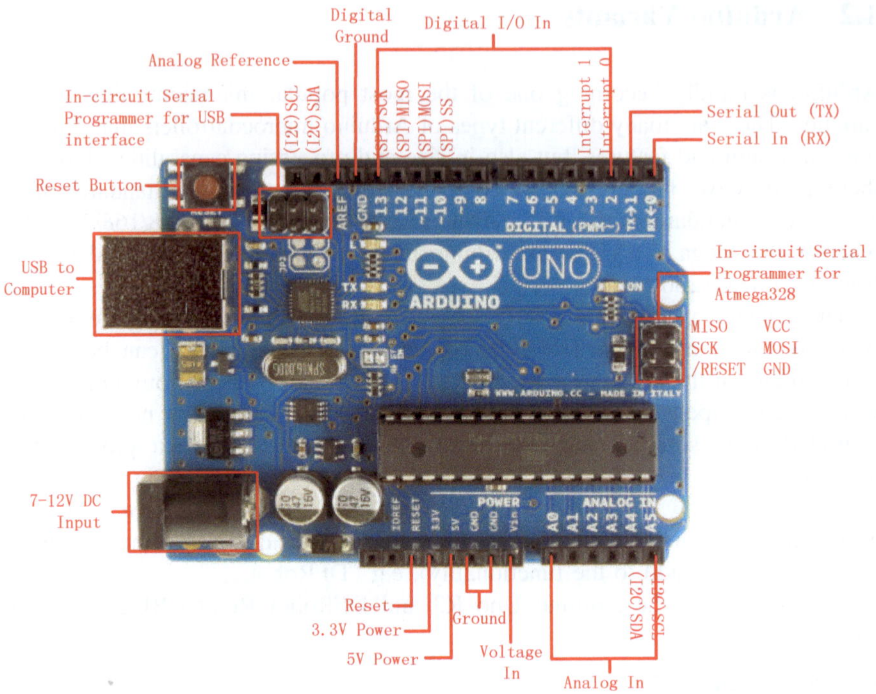

Fig. 1.1 Arduino UNO interface board

The Arduino Uno pinout is printed in the silkscreen in the top section. While this pinout is a good start, it is not a comprehensive guide. At first, you mainly use the pins in the female headers at the edge of the board (top and bottom in the photo), as well as the USB, and maybe the power

Tx and **Rx** are serial UART pins used for RS-232 and USB communications
I2C is another serial communications method that uses a bidirectional data line (SDA) and a clock line (SCL)
SPI is another serial communications method that uses one line for the master to transmit (MOSI—Master Out Slave In), another for the master to receive (MISO), and a third as a clock (SCK)
A/D, the Analog to Digital input, converts an analog voltage into a digital representation
PWM (Pulse Width Modulator) is used to create a square wave with a specific duty cycle (high time vs low time)
ICSP is the In Circuit Serial Programming—another way to program the processor
Vcc is the voltage supplied to the processor (+5VDC regulated from a higher input voltage)
3.3VDC is a regulated voltage (from the higher input voltage) for peripherals requiring that voltage—50 mA maximum

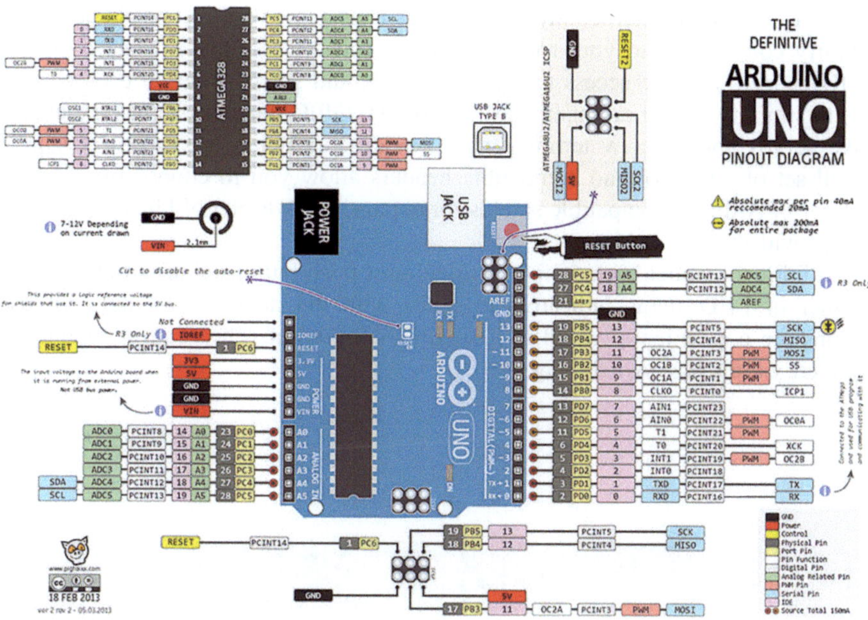

Fig. 1.2 Arduino Uno R3 pinout diagram

IOREF provides a voltage reference so shields can select the proper power source
AREF is a reference INPUT voltage used by the A/Ds
GND is the ground reference
RESET resets the processor (and some peripherals)

2. **DFRobot RoMeo BLE**

The **DFRobot Romeo BLE All-in-one Microcontroller (ATMega 328)** is an all-in-one Arduino-compatible microcontroller specifically designed for robotic applications. It benefits from the Arduino open-source platform; it is supported by thousands of open-source codes, and can easily be expanded with your Arduino-compatible shields.

This robot controller uses an Atmel ATmega328p as the main microcontroller. It comes preprogrammed with an Arduino bootloader for compatibility with the user-friendly Arduino Uno platform.

A secondary Texas Instruments CC2540 microcontroller handles the BLE Bluetooth Low Energy communication services. It comes preprogrammed with a firmware that supports transparent serial over Bluetooth and an AT command interpreter. Both microcontrollers are full programmable.

The Romeo robot controller contains a built in L298 dual channel motor driver chip. This motor driver can be used to drive two 5–23 V DC motors at up to 2 amps. Screw terminals are provided for connecting two motors and an external motor power supply.

The Romeo BLE microcontroller board also has a large number of pin headers to simplify connections to your robot project. A full set of 3-pin analog and digital GPIO headers provide access to signal, voltage, and ground lines at each connection to simplify wiring arrangements. The digital GPIO headers can also be used to drive servos and a screw terminal can provide an external servo power supply. A full set of Arduino Uno compatible headers allow you to choose from a large number of Arduino-compatible shields for expansion. A triplet of I2C connectors is also provided.

Additional features include five optional user push buttons, a reset button, and a number of LED status indicators to assist with diagnostics. The optional user buttons are conveniently wired to a single analog input. They can be enabled or disabled via a slider switch (Fig. 1.3).

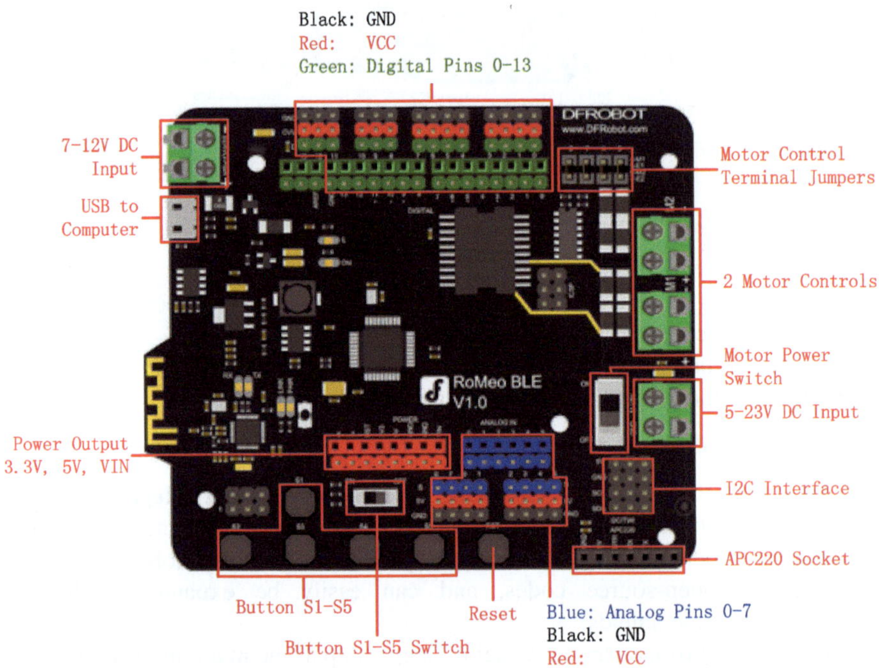

Fig. 1.3 Romeo BLE interface board

Specifications of the RoMeo BLE are as follows:

- Microcontroller: ATmega328P
- Bootloader: Arduino UNO
- Onboard BLE chip: TI CC2540
- 14 Digital I/O ports
- 6 PWM Outputs (Pin11, Pin10, Pin9, Pin6, Pin5, Pin3)
- 8 10-bit analog input ports
- 3 I2Cs
- Two way H-bridged motor driver with 2A maximum current
- 5 Buttons
- Power Supply Port: USB or DC2.1
- External Power Supply Range: 5–23 V
- DC output: 5 V/3.3 V
- Auto sensing/switching external power input
- Size: 94 mm × 80 mm

1.3 Install the Drivers

Before beginning your work, you must first download the development environment (the IDE) from here: www.arduino.cc/en/Main/Software.

1. Choose the right version for your operating system.
2. Download the file and double-click on it to open it; on Windows or Linux, this creates a folder named Arduino-[version], such as arduino-1.0.
3. Drag the folder to wherever you want it: your desktop, your Program Files folder (on Windows), etc. On the Mac, double-clicking it will open a disk image with an Arduino application (drag it to your Applications folder).
4. Whenever you want to run the Arduino IDE, you will need to open up the Arduino (Windows and Linux) or Applications folder (Mac), and double-click the Arduino icon. Do not do this yet, though; there is one more step.
5. You must install the drivers that allow your computer to communicate with your board through the USB port (Fig. 1.4).

Fig. 1.4 The dialog box of "Found New Hardware Wizard"

Plug the Arduino board into the computer; when the Found New Hardware Wizard window comes up. The **Found New Hardware Wizard** will open up as Windows will have detected that you have connected a new piece of hardware (your DFRobot Remeo board) to your PC. Tell it NOT to connect to Windows update (Select **No, not at this time**) and then click **Next**.

On the next page, select "**Install from a list or specific location (Advanced)**" and click **Next** (Fig. 1.5).

Make sure that "**Search for the best driver in these locations**" is checked. Uncheck "**Search removable media**." Check "**Include this location in the search**" and then click the Browse button. Browse to the location of the USB drivers and then click **Next** (Fig. 1.6).

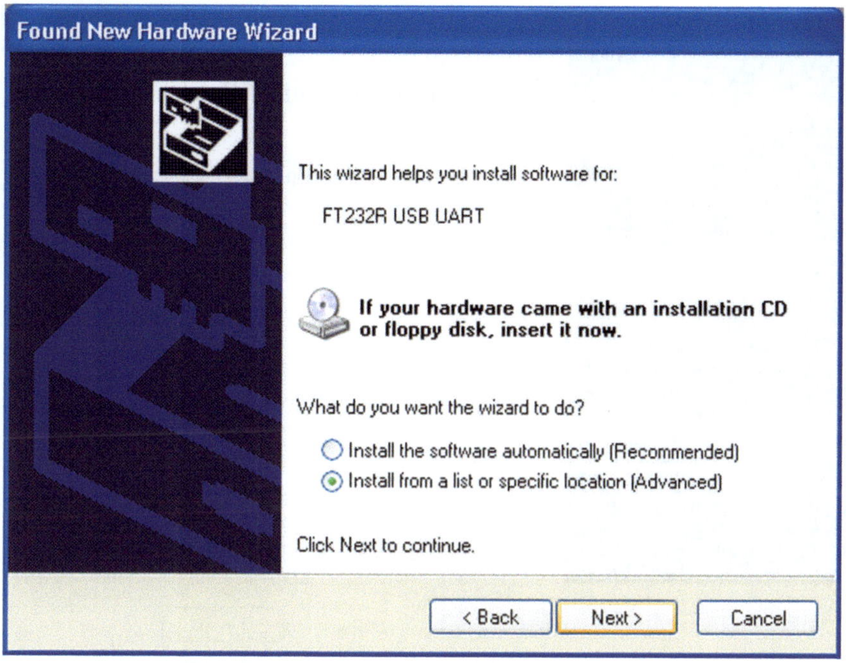

Fig. 1.5 Select "Install from a list or specific location (Advanced)"

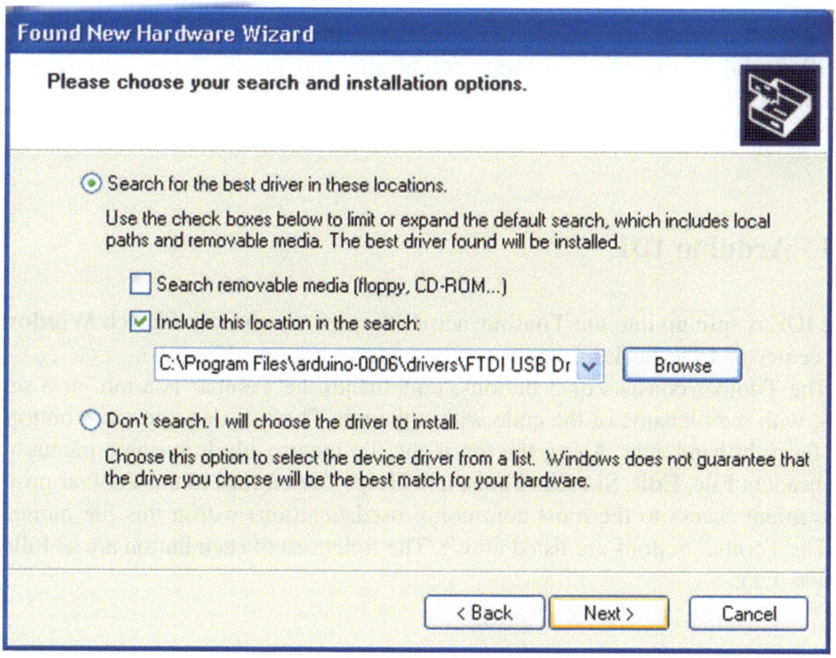

Fig. 1.6 Select "Search for the best driver in these locations"

Fig. 1.7 Completing the found new hardware wizard

The wizard will now search for a suitable driver and then inform you that a "**USB Serial Convertor**" has been found and that the hardware wizard is now complete. Click "**Finish**."

Now, you are ready to upload your first Sketch (Fig. 1.7).

1.4 Arduino IDE

The IDE is split up into the **Toolbar** across the top, the code, or **Sketch Window** in the center and the **Serial Output** window at the bottom (Fig. 1.8).

The Toolbar consists of 5 buttons, underneath the Toolbar is a tab, or a set of tabs, with the filename of the code within the tab. There is also one more button on the far right hand side. Along the top is the file menu with drop-down menus with the headers **File**, **Edit**, **Sketch**, **Tools,** and **Help**. The buttons in the Toolbar provide convenient access to the most commonly used functions within this file menu.

The Toolbar buttons are listed above. The functions of each button are as follows (Table 1.1):

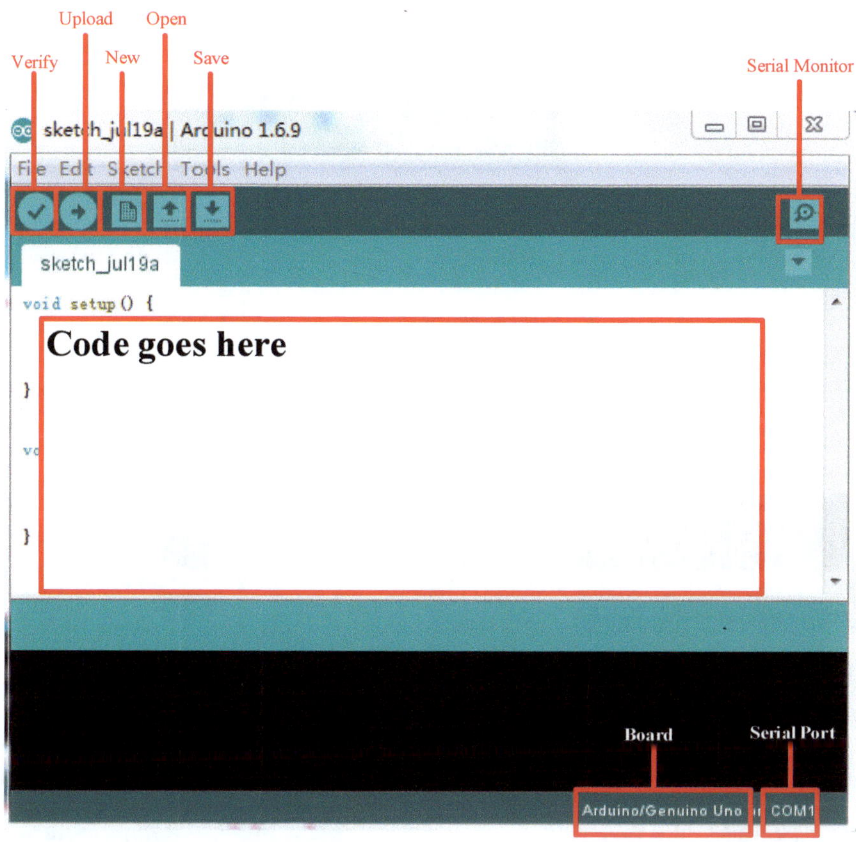

Fig. 1.8 Arduino IDE

Table 1.1 Toolbar buttons list

	Verify/compile	Checks the code for errors
	New	Creates a new blank Sketch
	Open	Shows a list of Sketches in your sketchbook
	Save	Saves the current Sketch
	Upload	Uploads the current Sketch to Arduino
	Serial Monitor	Displays serial data being sent from Arduino

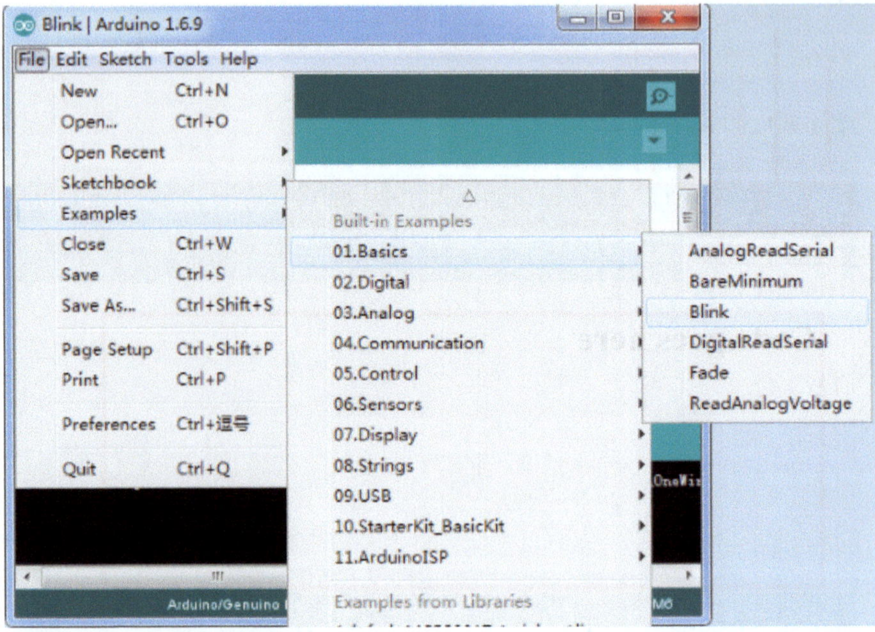

Fig. 1.9 "Blink" example in Arduino IDE

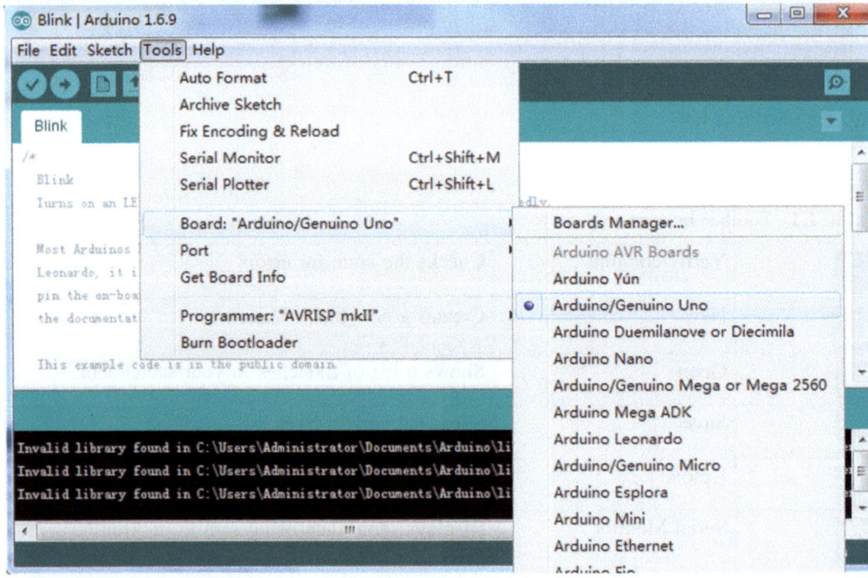

Fig. 1.10 Board type selection in Arduino IDE

Now, you are ready to test your first program with your Arduino board.

1. Open the **Blink** example sketch by going to: *File* → *Examples* → *01. Basics* → *Blink* (Fig. 1.9).
2. Select the type of Arduino board you are using: Tools → Board → your board type (Fig. 1.10).
3. Select the serial port that your Arduino is attached to: Tools → Port → COMxx (Fig. 1.11).
4. If you are not sure which serial device is your Arduino, have a look at the available ports, then unplug your Arduino and look again. The one that disappeared is your Arduino.
5. With your Arduino board connected, and the Blink sketch open, press the "Upload" button (Fig. 1.12).

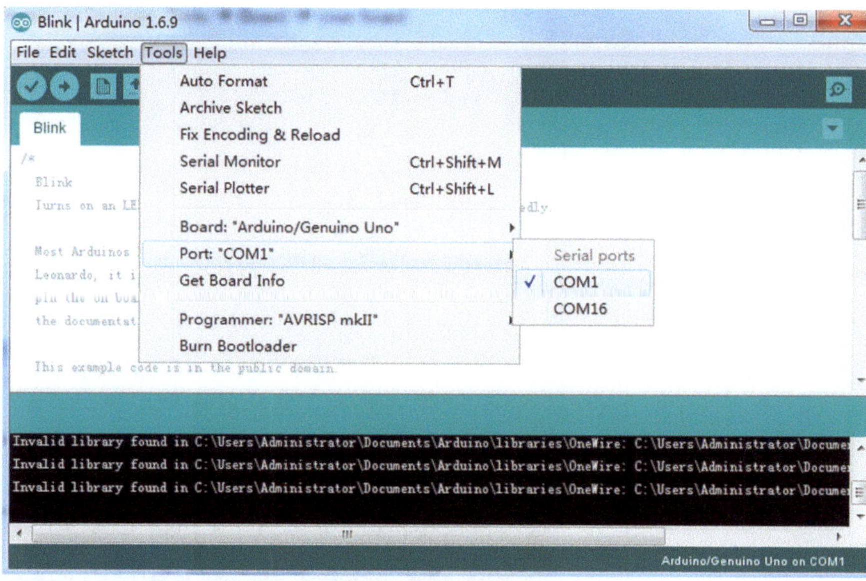

Fig. 1.11 Serial port selection in Arduino IDE

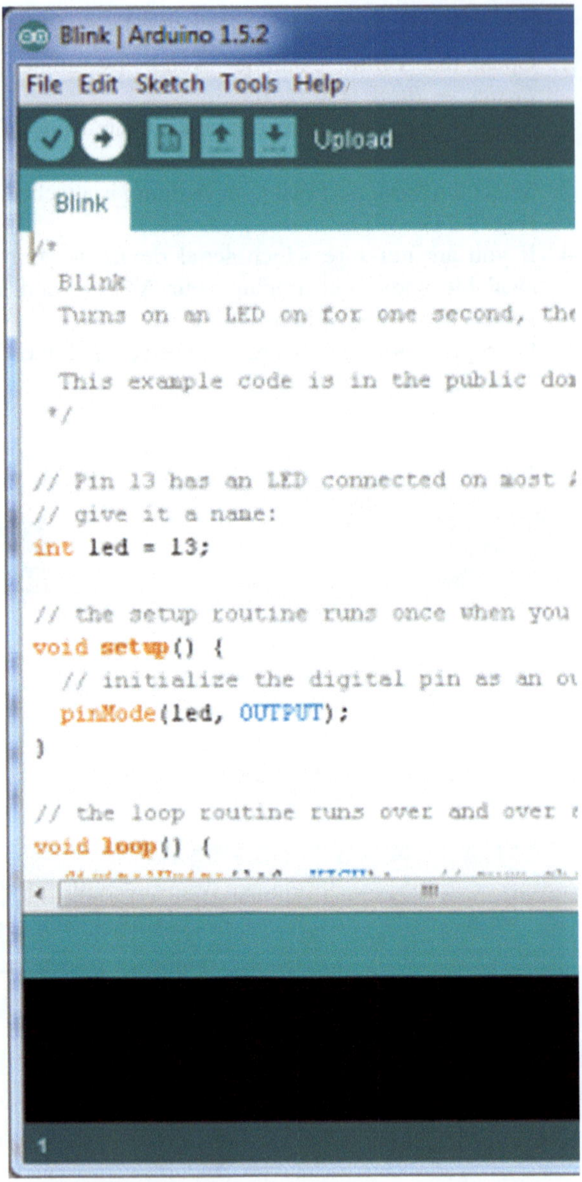

Fig. 1.12 Default codes of "Blink"

6. After a second, you should see some LEDs flashing on your Arduino, followed by the message "Done Uploading" in the status bar of the Blink sketch.
7. If everything worked, the onboard LED on your Arduino should now be blinking! You just programmed your first Arduino!

Chapter 2
The Basic Functions

2.1 Overview

The code you learn to write for your Arduino is very similar to the code you write in any other computer language. This implies that all the basic concepts remain the same and it is simply a matter of learning a new dialect. In the case of Arduino, the language is based on the C/C++ and can even be extended through C++ libraries. The IDE enables you to write a computer program, which is a set of step-by-step instructions that you then upload to the Arduino. Your Arduino will then carry out those instructions and interact with whatever you have connected to it. The Arduino includes many basic embedded functions, such as the functions for reading and writing to digital and analog input and output pins, interrupt functions, mathematical functions, and serial communication functions. Arduino functions are a convenient way to write code such as those for device drivers or commonly used utility functions. Furthermore, Arduino also consists of many built-in-examples. You just need to click on the toolbar menu: **File → Examples** to access them. These simple programs demonstrate all basic the Arduino commands. They span from a Sketch Bare Minimum, Digital, and Analog IO, to the use of Sensors and Displays.

For more information on the Arduino language, see the Language Reference section of the Arduino web site, http://arduino.cc/en/Reference/HomePage. All Arduino instructions are online.

2.2 Structure

The basic function of the Arduino programming language is fairly simple and runs in at least two parts. These two required parts, or functions, enclose blocks of statements.

T. Pan and Y. Zhu, *Designing Embedded Systems with Arduino*,
DOI 10.1007/978-981-10-4418-2_2

```
void setup() {
//code goes here
}
void loop() {
//code goes here
}
```

setup(): A function present in every Arduino sketch. Run once before the loop() function. The setup() function should follow the declaration of any variables at the very beginning of the program. It is the first function to run in the program, is run only once, and is used to set pinMode or initialize serial communication.

loop(): A function present in every single Arduino sketch. This code happens over and over again—reading inputs, triggering outputs, etc. The loop() is where (almost) everything happens and where the bulk of the work is performed.

2.3 Digital I/O Functions

Digital I/O will allow us to read the state of an input pin as well as produce a logical high or low at an output pin. If every potential external connection between a microcontroller and the outside world had a dedicated wire, the pin count for controller packages would be high. The ATmega 328P in the Romeo board has four 8-bit ports plus connections for power, ground and the like, yet it only has 28 physical pins. In general, each bit of a port can be programmed independently; some for input, some for output, or all of them for the same purpose.

1. **pinMode(pin, mode)**

Before we use a port, we need to inform the controller about how it should operate. In the Arduino system, this is usually done via a call to the library function **pinMode**(). Here is a description of the function from online references:

```
pinMode(pin,mode)
Parameters
   pin: the number of the pin whose mode you
        wish to set
   mode: INPUT, OUTPUT, or INPUT_PULLUP
Returns
   None
```

It should be noted that the pin could be a number or variable with a value ranging from 0 to 13 or A0 to A5 (when using the Analog Input pins for digital I/O) corresponding to the pin number printed on the interface board. Furthermore, Digital pins default as input, so you really only need to set them to OUTPUT in pinMode().

2. **digitalWrite(pin, value)**

Once a pin is established as an OUTPUT, it is then possible to turn that pin on or off using the digitalWrite() function. Its syntax is as follows:

```
digitalWrite(pin,value)
Parameters
   Pin: the number of the pin you want to write
   value: HIGH or LOW
Returns
   None
```

3. **digitalRead(pin)**

With a digital pin configured as an INPUT, we can read the state of that pin using the digitalRead() function. Its syntax is as follows:

```
digitalRead(pin)
Parameters
   pin: the number of the pin you want to read (int)
Returns
   HIGH or LOW
```

The pin can be specified as either a variable or constant (0–13) and the result is either HIGH or LOW.

4. **Example**

The following example reads a pushbutton connected to a digital input and turns on an LED connected to a digital output when the button is pressed. The circuit is shown in Fig. 2.1

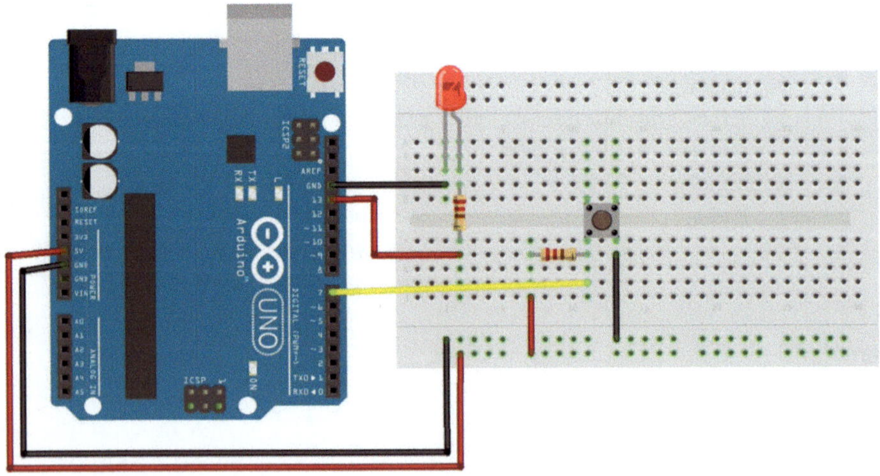

Fig. 2.1 Circuit layout for digital signal read and write

```
int led = 13;  // connect LED to pin 13
int pin = 7;   // connect pushbutton to pin 7
int value = 0; // variable to store the read value

void setup() {
  pinMode(led, OUTPUT);  // set pin 13 as output
  pinMode(pin, INPUT);   // set pin 7 as input
}
void loop() {
  value = digitalRead(pin); // set value equal to the pin 7 input
  digitalWrite(led, value); // set LED to the pushbutton value
}
```

2.4 Analog I/O Functions

1. **analogReference(type)**

The Arduino interface board, however, has a convenient pin called AREF located near digital pin 13 along with a function called analogReference() to provide the Arduino's ADC a reference voltage other than +5 V. This function will effectively increase the resolution available to analog inputs that operate at some other range of lower voltages below +5 V. The syntax for this function is as follows.

```
analogReference(type)
Parameters
   type: which type of reference to use (DEFAULT,
         INTERNAL, INTERNAL1V1, INTERNAL2V56, or
         EXTERNAL)
Returns
   None
```

DEFAULT: the default analog reference of 5 volts (on 5 V Arduino boards) or 3.3 volts (on 3.3 V Arduino boards)

INTERNAL: a built-in reference, equal to 1.1 volts on the ATmega168 or ATmega328 and 2.56 volts on the ATmega8 (not available on the Arduino Mega)

INTERNAL1V1: a built-in 1.1 V reference (Arduino Mega only)

INTERNAL2V56: a built-in 2.56 V reference (Arduino Mega only)
EXTERNAL: the voltage applied to the AREF pin (0–5 V only) is used as the reference.

The function is only called once in a sketch, but must be declared before analogRead() is used for the first time, making it a suitable candidate for placing in the setup() function. The type specified relates to the kind of reference voltage that we want to use.

2. **analogRead(pin)**

Reads the value from a specified analog pin with a 10-bit resolution. This function works with the above analogy only for pins (0–5). The analogRead() command will return a number including or between 0 and 1023.

```
analogRead(pin)
Parameters
   pin: the number of the analog input pin to read from
        (0-5)
Returns
   int(0 to 1023)
```

It takes about 100 μs (0.0001 s) to read an analog input, so the maximum reading rate is about 10,000 times a second. Furthermore, analogy pins unlike digital ones, do not need to be first declared as INPUT nor OUTPUT.

3. **analogWrite(pin,value)**

The Arduino also has the capability to output a Digital signal that acts as an Analog signal, this signal is called pulse width modulation (PWM). Digital Pins # 3, # 5, # 6, # 9, # 10, and # 11 have PWM capabilities. To output a PWM signal use the command: analogWrite().

```
analogWrite(pin,value)
Parameters
   pin: the number of the pin you want to write
   value: the duty cycle between 0 (always off, 0%) and
         255 (always on, 100%)
Returns
   None
```

You do not need to call pinMode() to set the pin as an output before calling analogWrite().

4. **Example**

The following example reads an analog value from an analogy input pin, converts the value by dividing by 4, and outputs a PWM signal on a PWM pin (Fig. 2.2).

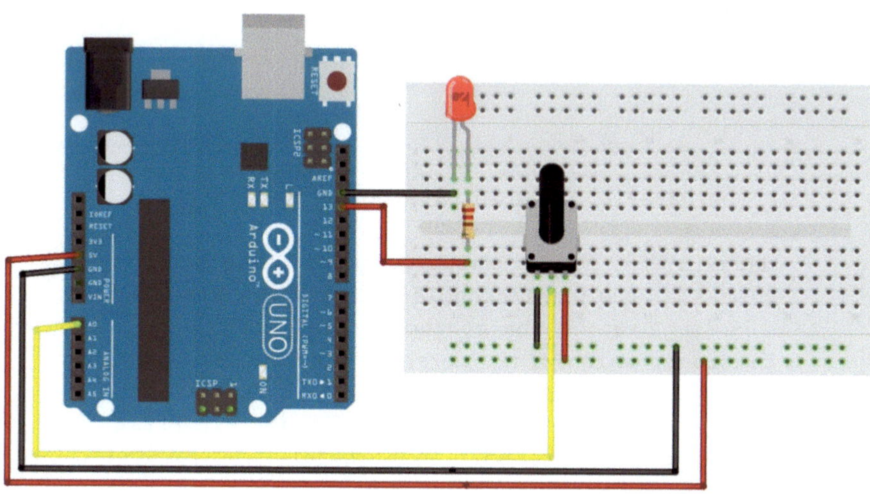

Fig. 2.2 Circuit layout for analogy signal read and write

```
int led = 13;  // connect LED to pin 13
int pin = 0;   // potentiometer on analogy pin 0
int value = 0; // variable to store the read value

void setup() {
}
void loop() {
  value = analogRead(pin); // set value equal to the pin 0's input
  value /= 4;              // converts 0-1023 to 0-255
  analogWrite(led, value); // output PWM signal to LED
}

```

2.5 Advanced I/O Functions

1. **shiftOut(dataPin,clockPin,bitOrder,value)**

Shifts out a byte of data one bit at a time. Starts from either the most (i.e., the leftmost) or least (rightmost) significant bit. Each bit is written in turn to a data pin, after which a clock pin is pulsed (taken high, then low) to indicate that the bit is available. The syntax is as follows.

```
shiftOut(dataPin,clockPin,bitOrder,value)
Parameters
   dataPin: the pin on which to output each bit (int)
   clockPin: the pin to toggle once the dataPin has been
             set to the correct value (int)
   bitOrder: which order to shift out the bits; either
             MSBFIRST or LSBFIRST. (Most Significant Bit
             First, or, Least Significant Bit First)
   value: the data to shift out (byte)
Returns
   None
```

This is known as synchronous serial protocol and is a common way that microcontrollers communicate with sensors, and with other microcontrollers. The two devices always stay synchronized, and communicate at close to maximum speeds, since they both share the same clock line. Often referred to as SPI (synchronous protocol interface) in hardware documentation.

2. **pulseIn(pin,value,timeout)**

Reads a pulse (either HIGH or LOW) on a pin. For example, if the value is HIGH, pulseIn() waits for the pin to go HIGH, starts timing, then waits for the pin to go LOW and stops timing. Returns the length of the pulse in microseconds. Gives up and returns 0 if no pulse starts within a specified time out. The timing of this function has been determined empirically and will probably show errors for longer pulses. Works on pulses from 10 μs to 3 min in length. The syntax is as follows.

```
pulseIn(pin,value,timeout)
Parameters
    pin: the number of the pin on which you want to read
         the pulse (int)
    value: type type of pulse to read: either HIGH or LOW
         (int)
    timeout (optional): the number of microseconds to wait
         for the pulse to start; default is one second
         (unsigned long)
Returns
    the length of the pulse (in microseconds) or 0 if no
         pulse started before the timeout
```

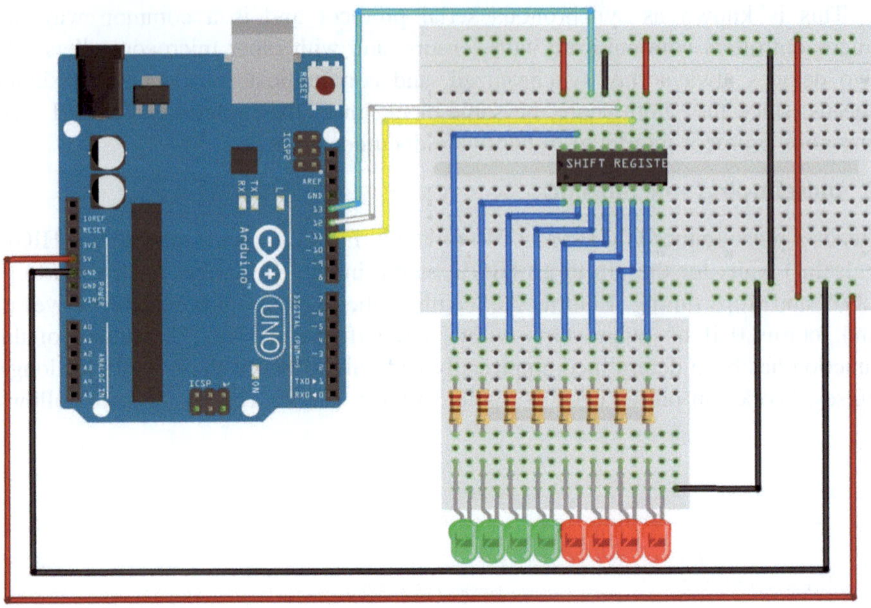

Fig. 2.3 Circuit layout for the shift register

Please go to Chap. 4, and refer the example for pulseIn().

3. **Example**

In this example, we will be using the 74HC595 8-bit shift register, which you can pick up from most places at a very reasonable price. This shift register will provide us with a total of eight extra pins to use. The layout is as follows (Fig. 2.3).

In this example, we increment the currentLED variable and pass it to the bitSet method. The bit is set to the left of the previous one to 1 every time, thereby informing the shift register to activate the output to the left of the previous one. As a result, the LEDs light up one by one.

```
int latchPin = 12;
int clockPin = 11;
int dataPin = 13;
byte leds = 0;
int currentLED = 0;
void setup() {
  pinMode(latchPin, OUTPUT);
  pinMode(dataPin, OUTPUT);
  pinMode(clockPin, OUTPUT);
  leds = 0;
}
void loop() {
  leds = 0;
  if (currentLED == 7) {
    currentLED = 0;
  }
  else {
    currentLED++;
  }
  bitSet(leds, currentLED);
  digitalWrite(latchPin, LOW);
  shiftOut(dataPin, clockPin, LSBFIRST, leds);
  digitalWrite(latchPin, HIGH);
  delay(250);
}
```

2.6 Timer Functions

1. **delay(ms)**

Pauses the program for the amount of time (in milliseconds) specified as the parameter. The syntax for the function is as follows.

```
delay(ms)
Parameters
    ms: the number of milliseconds to pause (unsigned
        long)
Returns
    None
```

Time is specified in milliseconds, where a delay of 1000 ms equals 1 s, 2000 ms equals 2 s, and so on. This value can be expressed as a constant or variable in the unsigned long data type.

The use of delay() in a sketch has significant drawbacks. No other reading of sensors, mathematical calculations, or pin manipulations can function during the delay function, so in effect, it brings most other activities to a halt.

2. **delayMicroseconds(us)**

Rather than a long delay, the delayMicroseconds() function is used to delay for a much shorter time. The syntax is as follows:

```
delayMicroseconds(us)
Parameters
    us: the number of microseconds to pause (unsigned int)
Returns
    None
```

Unlike delay(), time here is specified in microseconds, or millionths of a second, where a time period of 1000 μs would equal 1 ms or 0.001 of a second, 10,000 would equal 10 ms or 0.01 of a second, and so on.

3. **millis()**

Inside the microcontroller on the Arduino board there are three onboard hardware timers that work in the background to handle repetitive tasks like incrementing counters or keeping track of program operations. Each of these timers is already being used in some capacity, usually for handling hardware PWM and system timing. The millis() function makes use of one of these hardware timers to maintain a running counter of how many milliseconds the microcontroller has been running since the last time it was turned on or reset. Because this function uses a hardware timer, it performs its counting in the background with no impact on the flow or resources of our source code.

```
millis()
Parameters
    None
Returns
    Number of milliseconds since the program started
        (unsigned long)
```

By calling the function, it returns a value in milliseconds that can be used like any other variable as part of a conditional test, to perform arithmetic operations, or to be assigned to other variables. Because this function returns a value in an unsigned long data type, it will overflow, or reset to 0, in about 50 days. It can also result in undesired problems if an expression is performed on it using other data types like integers.

4. **micros()**

Where the millis() function returns the current operating time in milliseconds, the micros() function does the same, but in microseconds. This could be used in exactly the same manner as millis(), just on a much smaller scale, effectively returning the value 1000 for every 1 that millis() would return.

```
micros()
Parameters
    None
Returns
    Number of microseconds since the program started
        (unsigned long)
```

Unlike millis(), micros() will overflow, or reset back to 0, every 70 min.

2.7 Communication Functions

1. **Serial.begin(speed)**

Opens the serial port and sets the baud rate for serial data transmission. The typical baud rate for communicating with the computer is 9600, although other speeds are also supported, i.e., 300, 600, 1200, 2400, 4800, 9600, 14,400, 19,200, 28,800, 38,400, 57,600, or 115,200.

```
Serial.begin(speed)
Parameters
    speed: set the baud rate
Returns
    None
```

It should be noted that the digital pins 0 (RX) and 1 (TX) cannot be used at the same time when using serial communication.

2. **Serial.available()**

Receives the number of bytes (characters) available for reading from the serial port. This is data that has already arrived and been stored in the serial receive buffer.

```
Serial.available()
Parameters
    None
Returns
    the number of bytes available to read
```

Remember, the hardware serial port on the Arduino microcontroller has a buffer that can store up to 128 bytes of information so that it is not lost. If no data is waiting for us, it will return 0. On the other hand, if any data is available, the function will return a value other than 0, which will signify true. We can proceed to read from the buffer.

Serial.available() inherits from the Stream utility class.

3. **Serial.read()**

Reads incoming serial data.

```
Serial.read()
Parameters
    None
Returns
    the first byte of incoming serial data available (or
    -1 if no data is available) -int
```

This function simply returns the first byte of information available in the serial buffer. Because of the way our serial communications are structured, each character that we send to the Arduino through the Serial Monitor will be converted to that character's ASCII character value. For example, if we were to send the Arduino the number 1, instead of receiving the numerical integer 1, the Arduino will actually receive the numerical value 49 corresponding to that character's ASCII character code.

Serial.read() inherits from the Stream utility class

4. **Serial.print(val)**

Prints data to the serial port as human-readable ASCII text.

```
Serial.print(val)
Parameters
    val: the value to print - any data type
Returns
    None
```

This command can take many forms. Numbers are printed using an ASCII character for each digit. Floats are similarly printed as ASCII digits, defaulting to two decimal places. Bytes are sent as a single character.

5. **Serial.printIn(val,format)**

Prints data to the serial port as human-readable ASCII text followed by a carriage return character (ASCII 13, or '\r') and a newline character (ASCII 10, or '\n').

```
Serial.printIn(val,format)
Parameters
   val: the value to print - any data type
   format: specifies the number base (for integral data
           types) or number of decimal places (for
           floating point types)
Returns
   the number of bytes available to read
```

The println() function is a little easier to use and helps to clean up the output that we receive from the Serial Monitor. You will often see both the print() and println() functions used in conjunction to format the output, making the text easier to read.

6. **Example**

In this example, two Arduinos are used. The Arduino Uno on the left is our sender and the Arduino Mega on the right is our receiver. We use the Mega to make it easier to display debug information on the computer. The Arduinos are connected together using digitals 0 and 1 (RX and TX) on the Uno and digitals 16 and 17 (RX2 and TX2) on the Mega. The receiver on one needs to be connected to the transmit of the other, and vice versa. The Arduinos also need to have a common reference between the two. This is ensured by running a ground wire (Fig. 2.4).

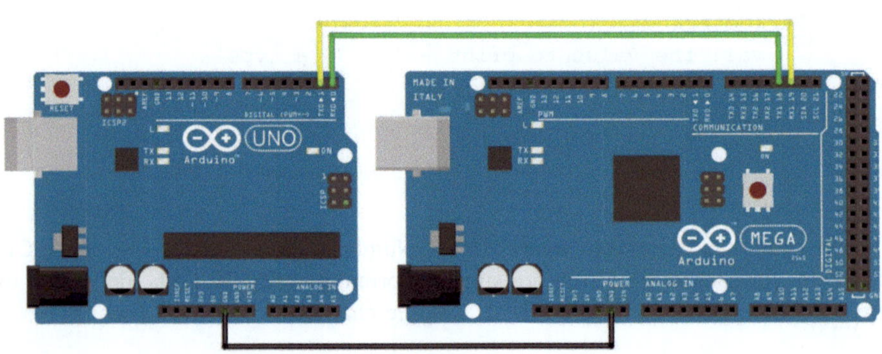

Fig. 2.4 The schematic of serial communication

The first step is to package the string to be communicated. In general, a packet is comprised of some start byte, a payload (the data you wish to send), and a checksum to validate your data. Here, the packet is: [0 × 53] + [counter value] + [static value] + [checksum].

Sender Code

The simple sender coder below increments our counter and sends our packet.

```
// Sender Information
unsigned char START_BYTE = 0x53; // ASCII "S"
unsigned char counterValue = 0;
unsigned char staticValue = 5;
unsigned char checksum = 0;

void setup() {
  Serial.begin(9600);
}
void loop() {
  // Increment our counter
  counterValue = counterValue + 1;
  // Check for overflow, and loop
  if (counterValue > 250)
    counterValue = 0;
  // Calculate our checksum
  checksum = counterValue + staticValue;
  // Important: Serial.write must be used, not print
  Serial.write(START_BYTE);
  Serial.write(counterValue);
  Serial.write(staticValue);
  Serial.write(checksum);
  // We only need to send a packet every 250ms.
  // If your code starts to get complicated,
  // consider using a timer instead of a delay
  delay(250);
}
```

Receiver Code

For the receiver code, we constantly go through the main loop and check whether we have information ready to be read. Once, we receive our first byte we compare it to our expected start byte. If this passes, then we set a flag and wait for the rest of the packet to roll in. Once, we have the expected packet then we read the values in it, calculate our checksum, and then print out the result on our terminal.

```
// Receiver Information
unsigned char START_BYTE = 0x53; // ASCII "S"
unsigned char counterValue = 0;
unsigned char staticValue = 0;
unsigned char checksum = 0;
boolean syncByteFound = 0; // Sync Byte flag

void setup() {
  Serial.begin(9600);
  Serial2.begin(9600);
}

void loop() {
  unsigned char rxByte = 0;
  unsigned char calculatedChecksum = 0;
  // Check to see if there's something to read
  if (Serial2.available() > 0) {
    // If we're waiting for a new packet, check for the sync byte
    if (syncByteFound == 0) {
      rxByte = Serial2.read();
      if (rxByte == 0x53)
        syncByteFound = 1;
    }
    // If we've found our sync byte, check for expected number of
bytes
    if (Serial2.available() > 2) {
      counterValue = Serial2.read();
      staticValue = Serial2.read();
      checksum = Serial2.read();
      calculatedChecksum = counterValue + staticValue;
      // Print out our serial information to debug
      Serial.print("[");
      Serial.print("S");
      Serial.print("]");
      Serial.print("[");
      Serial.print(counterValue);
      Serial.print("]");
      Serial.print("[");
      Serial.print(staticValue);
      Serial.print("]");
      Serial.print("[");
```

```
            Serial.print(checksum);
            Serial.print("]");
            if (calculatedChecksum == checksum)
              Serial.println("[Checksum Passed]");
            else
              Serial.println("[Checksum FAILED]");
            syncByteFound = 0;
        }
    }
}
```

2.8 Interrupt Functions

1. attachInterrupt(digitalPinToInterrupt(pin),ISR,mode)

The attachInterrupt() function enables hardware interrupts and links a hardware pin to an ISR to be called when the interrupt is triggered. This function also specifies the type of state change that will trigger the interrupt. Its syntax is as follows:

```
attachInterrupt(digitalPinToInterrupt(pin),ISR,mode)
Parameters
    interrupt: the number of the interrupt (int)
    pin: the pin number
    ISR: the interrupt service routine (ISR) to call when the
         interrupt occurs; this function must take no
         parameters and return nothing. This function is
         sometimes referred to as an interrupt service routine.
    mode: defines when the interrupt should be triggered. Four
         constants are predefined as valid values:
       LOW to trigger the interrupt whenever the pin is low,
       CHANGE to trigger the interrupt whenever the pin changes
         value
       RISING to trigger when the pin goes from low to high,
       FALLING for when the pin goes from high to low.
Returns
    None
```

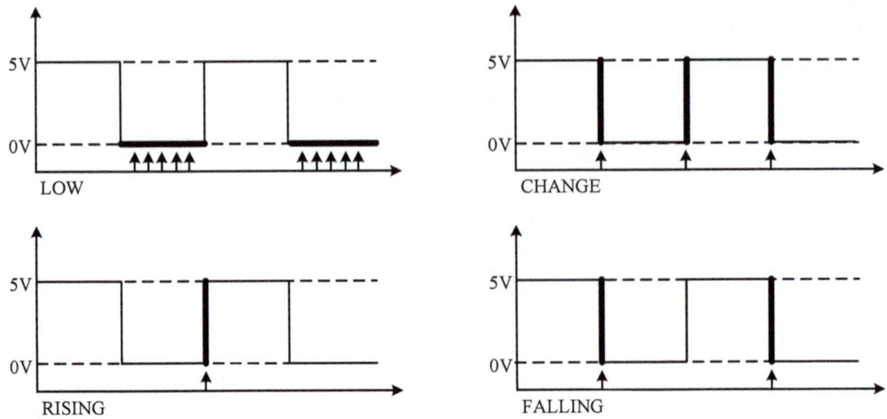

Fig. 2.5 State changes

Normally, you should use digitalPinToInterrupt(pin), rather than place an interrupt number directly into your sketch. The specific pins with interrupts, and their mapping to interrupt numbers vary for each type of board. On the DFrobot Romeo board, there are two possible hardware interrupts, 0 (for digital pin 2) and 1 (for digital pin 3).

Four possible modes are shown in Fig. 2.3, which include LOW, CHANGE, RISING, and FALLING (Fig. 2.5).

2. **detachInterrupt(interrupt)**

In some cases, we might need to change the mode of an enabled interrupt. For example, we may change the mode from RISING to FALLING. For this, we need to first stop the interrupt by using the detachInterrupt() function. Its syntax is as follows:

```
detachInterrupt(interrupt)
Parameters
    interrupt: the number of the interrupt to disable
Returns
    None
```

With only one parameter to determine which interrupt we are disabling, this parameter is specified as either 0 or 1. Once the interrupt has been disabled, we can then reconfigure it using a different mode in the attachInterrupt() function.

3. **interrupts()**

All interrupts in Arduino can be enabled by the function interrupts(). The syntax is as follows:

```
Interrupts()
Parameters
    None
Returns
    None
```

Interrupts allow certain important tasks to run in the background and are enabled by default. Some functions will not work while interrupts are disabled, and incoming communication may be ignored. Interrupts can slightly disrupt the timing of a code, however, and may be disabled for particularly critical sections of code.

4. **noInterrupts()**

To deactivate all interrupts, we can use the function noInterrupts(). The syntax is as follows.

```
noInterrupts()
Parameters
    None
Returns
    None
```

5. **Example**

In this example, we blink the built-in LED every 500 ms, during which time both interrupt pins are monitored. When the button on the interrupt 0 is pressed, the value for micros() is displayed on the Serial Monitor, and when the button on the interrupt 1 is pressed, the value for millis() is displayed (Fig. 2.6).

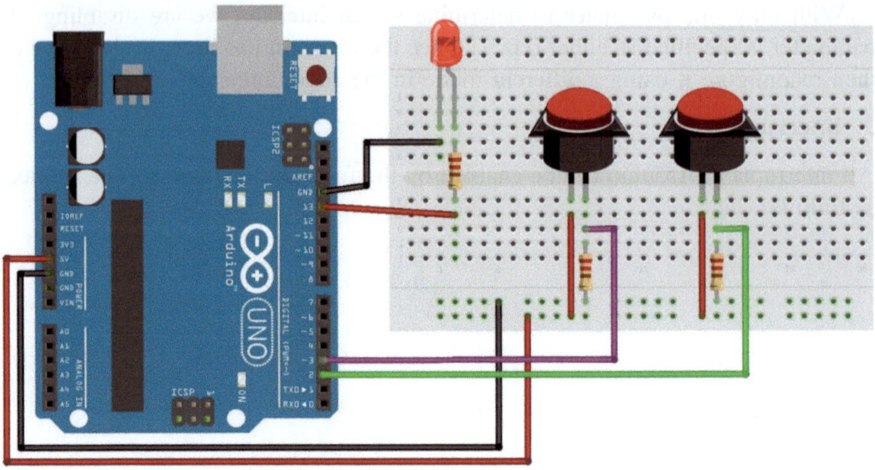

Fig. 2.6 The circuit for interrupts and time functions

```
#define LED 13
void setup() {
  Serial.begin(9600);
  pinMode(LED, OUTPUT);
  attachInterrupt(0, displayMicros, RISING);
  attachInterrupt(1, displayMillis, RISING);
}
void loop() {
  digitalWrite(LED, HIGH);
  delay(500);
  digitalWrite(LED, LOW);
  delay(500);
}
void displayMicros() {
  Serial.write("micros()=");
  Serial.println(micros());
}
void displayMillis() {
  Serial.write("millis()=");
  Serial.println(millis());
}
```

2.9 Math Functions

The Arduino language supports many math functions which make it possible to perform sophisticated data processing in a data logger sketch that would otherwise have to be performed offline in some other software application.

1. **min(x,y)**

   Calculates the minimum of two numbers

```
min(x,y)
Parameters
    x: the first number, any data type
    y: the second number, any data type
Returns
    The smaller of the two numbers
```

2. **max(x,y)**

Calculates the maximum of two numbers.

```
max(x,y)
Parameters
    x: the first number, any data type
    y: the second number, any data type
Returns
    The larger of the two numbers
```

3. **abs(x)**

Computes the absolute value of a number

```
abs(x)
Parameters
    x: the number
Returns
    x: if x is greater than or equal to 0.
    -x: if x is less than 0.
```

Because of the way the abs() function is implemented, avoid using other functions inside the brackets, it may lead to incorrect results.

4. **Trigonometric functions**

The trigonometric functions include sin(rad), cos(rad), and tan(rad). All trigonometric functions accept as input and return as output angles in radians, not degrees:radians = degrees $\times \pi/180$ and vice versa to convert radians back to degrees.

5. **pow(base,exponent)**

Calculates the value of a number raised to a power. pow() can be used to raise a number to a fractional power. This is useful for generating exponential mapping of values or curves.

```
pow(base,exponent)
Parameters
    base: the number (float)
    exponent: the power to which the base is raised (float)
Returns
    The result of the exponentiation (double)
```

6. **sqrt(x)**

Calculates the square root of a number.

```
sqrt(x)
Parameters
    x: the number, any data type
Returns
    double, the number's square root
```

7. **constrain(x,a,b)**

Constrains a number to be within a range.

```
constrain(x,a,b)
Parameters
    x: the number to constrain, all data types
    a: the lower end of the range, all data types
    b: the upper end of the range, all data types
Returns
    x: if x is between a and b
    a: if x is less than a
    b: if x is greater than b
```

8. **map(value,fromLow,fromHigh,toLow,toHigh)**

Remaps a number from one range to another. That is, a value of fromLow would be mapped to toLow, a value of fromHigh to toHigh, values in-between to values in-between, etc.

```
map(value,fromLow,fromHigh,toLow,toHigh)
Parameters
    value: the number to map
    fromLow: the lower bound of the value's current range
    fromHigh: the upper bound of the value's current range
    toLow: the lower bound of the value's target range
    toHigh: the upper bound of the value's target range
Returns
    The mapped value
```

9. **random(min,max)**

The random() function returns a semi-random number up to the parameters specified. If no parameters are specified, it will return a value in the signed long data type, with a range of −2,147,483,648–2,147,483,647. Its syntax is as follows:

```
random(min,max)
Parameters
    min: lower bound of the random value, inclusive (optional)
    max: upper bound of the random value, exclusive
Returns
    a random number between min and max
```

The random() function will assume a value of 0 as its minimum value.

10. **Example**

In this example, we create a sine wave and configure the brightness of the LED to follow the path of the wave. This is what makes the light pulsate in the form of a sine wave instead of just illuminating up to full brightness and back down again (Fig. 2.7).

The codes are as follows (Fig. 2.7)

```
int ledPin = 11;
float sinVal;
int ledVal;
void setup() {
  pinMode(ledPin, OUTPUT);
}
void loop() {
  for (int x = 0; x < 180; x++) {
    // convert degrees to radians
    // then obtain sin value
    sinVal = (sin(x * (3.1412 / 180)));
    ledVal = int(sinVal * 255);
    analogWrite(ledPin, ledVal);
    delay(25);
  }
}
```

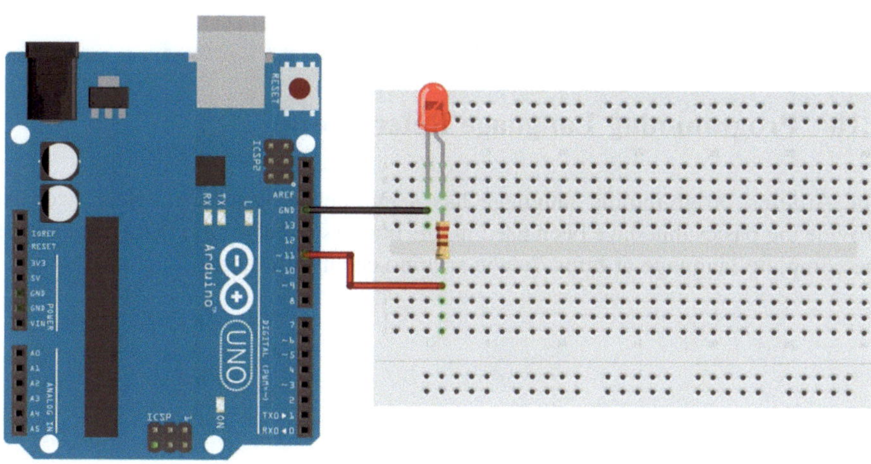

Fig. 2.7 The circuit for mathematical example

Table 2.1 Programming language reference

Construct	Description
int	Integer values, such as 123
float	Decimal values, such as 1.15
char[]	String values, such as "Arduino"
HIGH	Digital pin with current
LOW	Digital pin with no current
INPUT	Pin can only be read
OUTPUT	Pin can only be set
A0–A7	Constants for analog pins; varies by board
0–13	Value for digital pins; varies by board
analogRead()	Returns analog pin value (0–1023)
analogWrite()	Sets analog pin value
digitalRead()	Returns digital pin value (HIGH or LOW)
digitalWrite()	Sets digital pin value (HIGH or LOW)
Serial.begin()	Initializes serial monitor
Serial.print()	Logs message on serial monitor
Serial.println()	Logs message on serial monitor with new line
delay(ms)	Adds a wait in processing
setup()	Standard Arduino function called once
loop()	Standard Arduino function called repeatedly
if	Checks for a true/false condition
if … else	Checks for a true/false condition; if false goes to else
//	Single-line comment
/* */	Multiline comment
#define	Defines a constant
#include	Includes an external library

2.10 Programming Language Reference

The Arduino programming language has a number of constructs. Here, we just provide the basics that have been used in this book (see Table 2.1). You can explore the complete language at https://www.arduino.cc/en/Reference.

Chapter 3
Using Sensors with the Arduino

3.1 Introduction

The definition of a sensor is "a device that senses a variation in input energy to produce a variation in another or the same form of energy." In general, the sensing principles are physical or chemical in nature and they can be grouped according to the form of energy in which the signals are received and generated. There are six types of signals on the basis of the energy generated or received and they are mechanical, thermal, electrical, magnetic, radiant, and chemical. Here, we focus on sensors that are readily available and offer the most versatility for the price.

3.2 Light Sensitive Sensors

3.2.1 Introduction

Light sensitive sensors are electromagnetic radiation detectors that function in the ultraviolet to far infrared spectral range. The electric signal of a light sensor can be produced by either a quantum or thermal response from a sensing material when photons are absorbed by such material. Therefore, light sensors can be divided into two major groups, quantum and thermal. Light sensors rely on the interaction of individual photons with a crystalline lattice of semiconductor materials. Their operations are based on the photo-effect, which requires, at the least, the energy of a single photon concentrated into localized bundles under certain circumstances. The energy of a single photon is given by

$$E = h\nu \tag{3.1}$$

T. Pan and Y. Zhu, *Designing Embedded Systems with Arduino*,
DOI 10.1007/978-981-10-4418-2_3

where v is the frequency of light, and h is Planck's constant ($h = 6.626 \times 10^{-34}$ JS). When a photon strikes a conductor, it may result in the generation of a free electron. The photoelectric effect is shown as

$$hv = \phi + K_{\mathrm{m}} \tag{3.2}$$

where ϕ is the work function of photon energy E, which is used to detach the electron from the surface. K_{m} is the maximum kinetic energy of the electron upon it exiting the surface. The electron becomes mobile when the photon transfers its high energy to an electron. Such a response results in an electric current.

3.2.2 *Photodiodes*

A photodiode is a reverse-biased semiconducting optical sensor that is biased against its easy flow direction so that the current is very low. Such a device has a band structure (such as a P-N junction) in which the permitted energies in the structure change. In a photodiode, an electron is freed when a photon is absorbed, it may pass over the energy barrier if it possesses enough energy. In this respect, the photodiode only produces a current if the absorbed photon has more energy than that needed to traverse the P-N junction. The photodiode is said to have a cutoff wavelength because the lesser the wavelength of light responses the greater the energy. Therefore, a photodiode produces detectable currents for photons with wavelength less than the cutoff, while a current is not produced if the wavelengths of the photons are greater than the cutoff.

The schematic of a photodiode is shown in Fig. 3.1. The voltage of the PIN port changes when Q1 absorbs photons with wavelengths greater than the cutoff. Else, Q1 is not available for the current conduction.

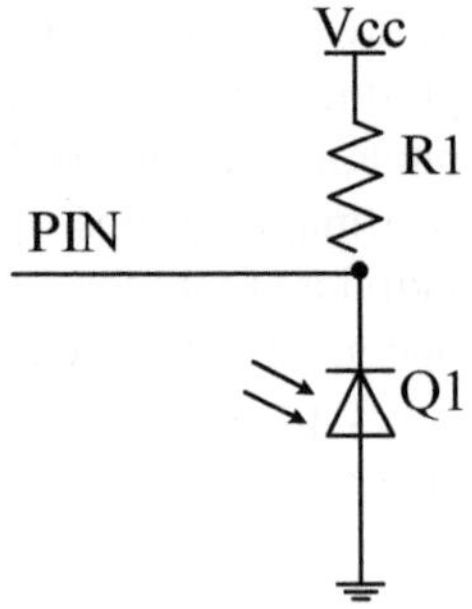

Fig. 3.1 Schematic of a photodiode

3.2.3 *Demonstration*

1. **Components**

   - DFRobot UNO R3 board and USB cable × 1.
   - DFR0026 (analog ambient light sensor) × 1.
   - LED × 1.
   - Resistor (220 Ω) × 1.
   - Jumper wires × *n*.

2. **Hardware Setting**

DFR0026 has three pins: VCC, Output, and GND. The VCC should be connected to 5 V and the GND to a common ground with your Arduino. The Output pin of DFR0026 should be plugged into a defined pin on the DFRobot UNO R3 board (here, analog input PIN 0) (Fig. 3.2).

3. **Sample Codes**

```
int LED = 13; //define LED digital pin 13
int LIGHT = 0; //define light analog pin 0
int val = 0; //define the voltage value
void setup() {
  pinMode(LED, OUTPUT); //Configure LED as output mode
  Serial.begin(9600); //Configure baud rate 9600
}
void loop() {
  val = analogRead(LIGHT); // Read voltage value (0 - 1023)
  Serial.println(val); // read voltage value from serial monitor
  if (val < 700) { // If lower than 700, turn off LED
    digitalWrite(LED, LOW);
  }
  else { // Otherwise turn on LED
    digitalWrite(LED, HIGH);
  }
  delay(10); // delay for 10ms
}
```

4. **Results**

After uploading the code, you can shine a flashlight on the photodiode to alter the light levels in the environment. When it is dark, the LED should light up. When it is bright, the LED should turn off.

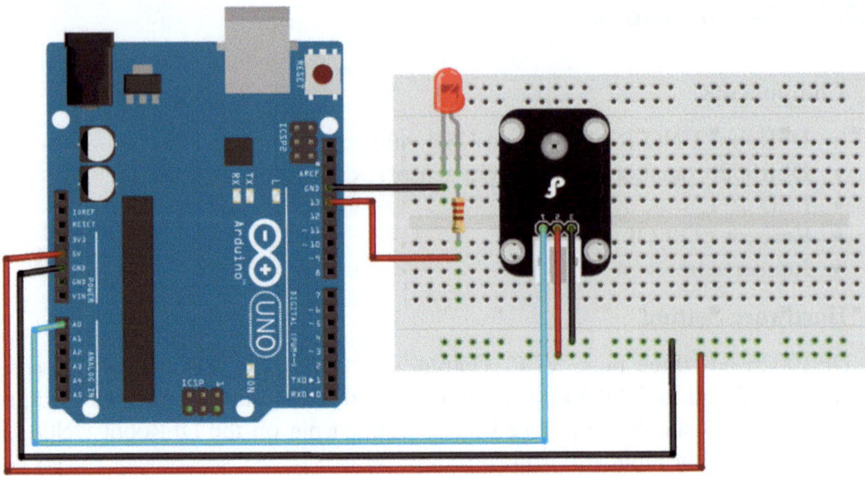

Fig. 3.2 A diagram of the layout of the DFR0026 and UNO R3

Furthermore, you can open the serial monitor (Baudrate = 9600) and see what outputs the photodiode provides. Then, use the number you receive as a comparison number to alter the sensitivity of the circuit (Fig. 3.3).

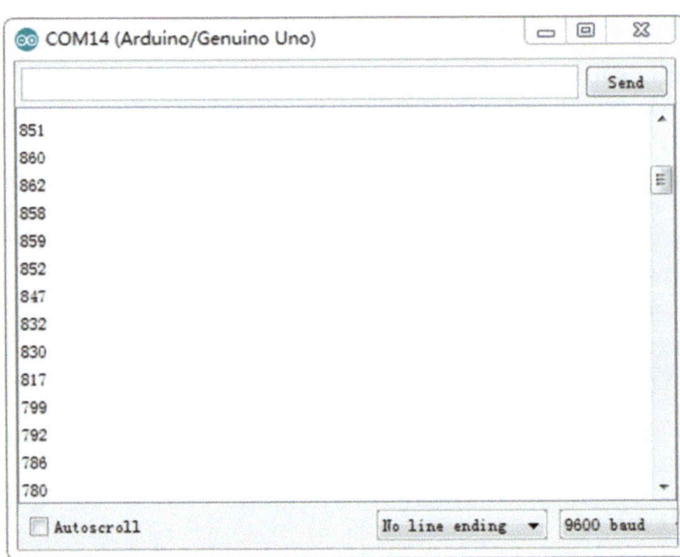

Fig. 3.3 Log messages from Arduino using DFR0026

3.3 Temperature Sensors

3.3.1 Introduction

In many systems, temperature control is fundamental. There are a number of passive and active temperature sensors that can be used to measure system temperature, including: thermocouple, resistive temperature detector, thermistor, and silicon temperature sensors. These sensors provide temperature feedback to a system controller to make decisions such as, over-temperature shutdown, turn-on/off cooling fan, temperature compensation, or general purpose temperature monitoring.

There are many kinds of thermal management products, including Logic Output, Voltage Output, and Serial Output Temperature Sensors. These products allow a system designer to implement a device that best meets their application's requirements. Key features include high accuracy, low power, extended temperature range, and small packages.

3.3.2 Digital Temperature Sensor

3.3.2.1 Introduction

DS18B20 is a direct-to-digital temperature sensor. It has an operating temperature range of −55 to +125 °C and is accurate up to ±0.5 °C over the range of −10 to +85 °C. Further, it provides 9–12 bit Celsius temperature measurements and has an alarm function with nonvolatile user programmable upper and lower trigger points. DS18B20 communicates over a 1-Wire bus that by definition requires only one data line (and ground) for communication with a central microprocessor. Each DS18B20 has a unique 64-bit serial code, which allows multiple DS18B20s to function on the same 1-Wire bus. Thus, it is simple to use one microprocessor to control many DS18B20s distributed over a large area. Applications that can benefit from this feature include HVAC environmental controls, temperature monitoring systems inside buildings, equipment or machinery, and process monitoring and control systems.

The block diagram of DS18B20 is shown in Fig. 3.4. The sensor can be powered by an external supply on the VCC pin, or it can operate in "parasite power" mode, which allows the DS18B20 to function without a local external supply. Parasite power is very useful for applications that require remote temperature sensing or that which are very space constrained. The advantage of the external power supply is that the MOSFET pull-up is not required, and the 1-Wire bus is free to carry other traffic during the temperature conversion time.

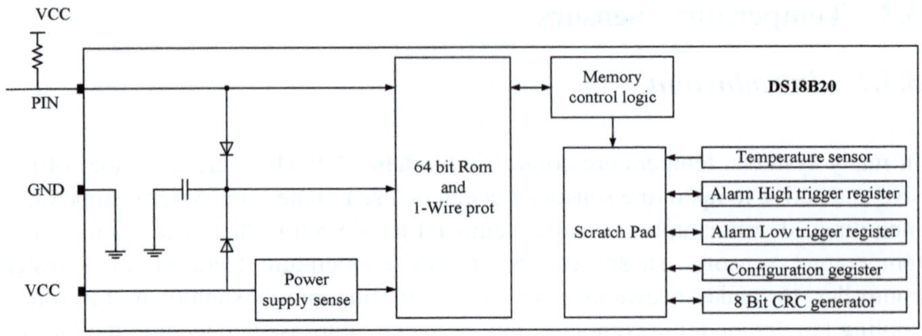

Fig. 3.4 Block diagram of DS18B20

8-BIT CRC	48-BIT SERIAL NUMBER	8-BIT FAMILY CODE(28h)
MSB LSB	MSB LSB	MSB LSB

Fig. 3.5 Unique 64-bit code of DS18B20

The unique 64-bit code stored in the ROM is shown in Fig. 3.5. Where the least significant 8 bits of the ROM code contain the DS18B20's 1-Wire family code: 28 h. The next 48 bits contain a unique serial number. The most significant 8 bits contain a cyclic redundancy check (CRC) byte that is calculated from the first 56 bits of the ROM code.

3.3.2.2 Demonstration

1. **Components**
   - DFRobot UNO R3 board and USB cable × 1.
   - DFR0024 (DS18B20) × 1.
   - DFR0032 (digital buzzer) × 1.
   - Jumper wires × *n*.
2. **Hardware Setting**

DFR0024 (18B20) is used here to measure the environmental temperature. The sensor has three pins: VCC, Output, and GND. The VCC should be connected to 5 V and the GND to a common ground with your Arduino. The Output pin of DFR0024 should be plugged into a defined pin on the DFRobot UNO R3 board (here, PIN 3).

DFR0032 has three pins: VCC, Output, and GND. The Output pin should be plugged into a defined pin on the DFRobot UNO R3 board (here, PIN 4) (Fig. 3.6).

3. **Sample Codes**

Before we verify these codes with Arduino, the 1-wire library, i.e., "OneWire.h", should be downloaded first,

(http://www.pjrc.com/teensy/arduino_libraries/OneWire.zip). Then, import the library in the Arduino IDE with the path as "Sketch → Include Library → Add.Zip Library."

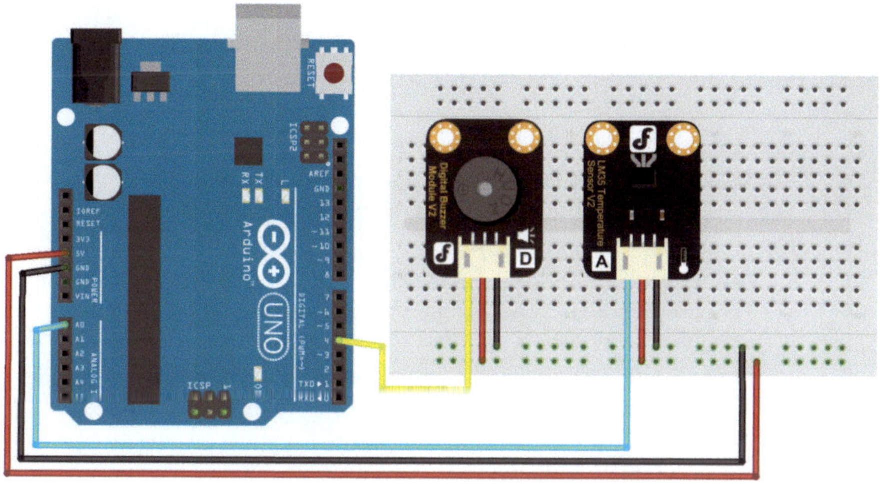

Fig. 3.6 A diagram of the layout of the DFR0024, DFR0032 and UNO R3

```
#include <OneWire.h>
int ALARM = 4; //define Buzzer digital pin 4
int DS18S20 = 3; //DS18S20 Signal pin on digital 3
float tmprVal = 0; //define value
float sinVal;
int toneVal;
unsigned long tepTimer;
OneWire ds(DS18S20); // on digital pin 3

void setup(void) {
  Serial.begin(9600); // configure baud rate to 9600 bps
  pinMode(ALARM, OUTPUT); // configure pin of buzzer
}

void loop(void) {
  tmprVal = getTemp();
  if (tmprVal > 27) { //If temperature > 27, the buzzer starts to
make sound.
    for (int x = 0; x < 180; x++) {
      //Convert sin function to radian
      sinVal = (sin(x * (3.1412 / 180)));
      //Use sin function to generate frequency of sound
      toneVal = 2000 + (int(sinVal * 1000));
      //Configure the buzzer pin 4
      tone(ALARM, toneVal);
      delay(2);
    }
  }
  else { // If the temperature <= 27, turn off the buzzer
    noTone(ALARM); // Turn off the buzzer
  }
  if (millis() - tepTimer > 50) { // Every 500 ms, serial port
outputs temperature value
    tepTimer = millis();
    Serial.print("temperature: ");
    Serial.print(tmprVal);
    Serial.println("C");
  }
}
```

```

//returns the temperature from one DS18S20 in DEG Celsius
float getTemp() {
  byte data[12];
  byte addr[8];
  if (!ds.search(addr)) {
    //no more sensors on chain, reset search
    ds.reset_search();
    return -1000;
  }
  if (OneWire::crc8(addr, 7) != addr[7]) {
    Serial.println("CRC is not valid!");
    return -1000;
  }
  if (addr[0] != 0x10 && addr[0] != 0x28) {
    Serial.print("Device is not recognized");
    return -1000;
  }
  ds.reset();
  ds.select(addr);
  ds.write(0x44, 1); // start conversion, with parasite power on
at the end
  byte present = ds.reset();
  ds.select(addr);
  ds.write(0xBE); // Read Scratchpad
  for (int i = 0; i < 9; i++) { // we need 9 bytes
    data[i] = ds.read();
  }
  ds.reset_search();
  byte MSB = data[1];
  byte LSB = data[0];
  float tempRead = ((MSB << 8) | LSB); //using two's compliment
  float TemperatureSum = tempRead / 16;
  return TemperatureSum;
}
```

4. **Results**

Read temperature values from the serial port. If you place your fingers on the DFR0024 sensor, you will find that the temperature rises immediately. Your fingers are transferring heat to the sensor (Fig. 3.7).

Once the temperature reaches 27 °C, the buzzer starts to sound. If the temperature drops below 27 °C, the buzzer stops.

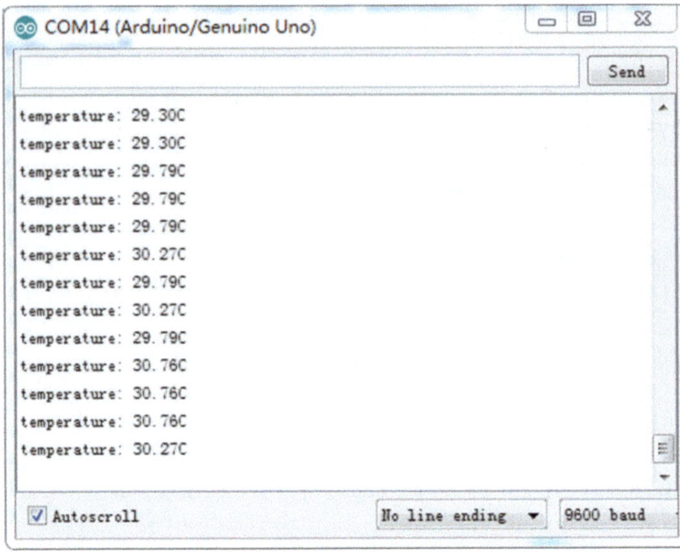

Fig. 3.7 Log messages from Arduino using DFR0024

3.3.3 Analog Temperature Sensor

3.3.3.1 Introduction

The analog sensor acts as a variable resistor. As the temperature increases, the voltage output of the sensor decreases. Once we measure the voltage output, we can calibrate the sensor and convert the output voltage to temperature. The analog sensor includes thermocouples, platinum resistance, thermal resistance, and temperature semiconductor chips, which are commonly used in high temperature measurement thermocouples. A platinum resistance temperature is used in the measurement of 800 °C, while a thermal resistance and semiconductor temperature sensor is suitable for measuring temperatures of 100–200 °C or below, in which the application of a simple semiconductor temperature sensor has good linearity and high sensitivity. One of the semiconductor temperature sensors is the LM35 series. The LM35 series are precision integrated-circuit temperature sensors, whose output voltage is linearly proportional to the Celsius temperature. The LM35, thus, has an advantage over linear temperature sensors calibrated in Kelvin, as the user is not required to subtract a large constant voltage from its output to obtain a convenient Centigrade scaling.

The high-accuracy version of the LM35 does not require any external calibration or trimming to provide typical accuracies of ±1/4 °C at room temperature and ±3/4 °C over a full −55 to +150 °C temperature range. The typical application is shown in Fig. 3.8.

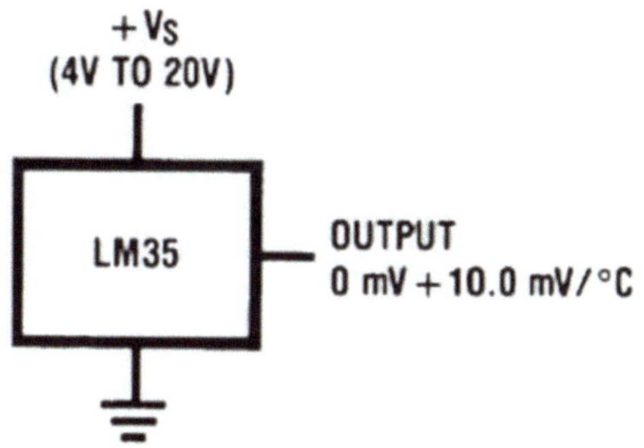

Fig. 3.8 Basic centigrade temperature sensor

3.3.3.2 Demonstration

1. **Components**
   - DFRobot UNO R3 board and USB cable × 1.
   - DFR0023 (LM35) × 1.
   - DFR0032 (digital buzzer) × 1.
   - Jumper wires × *n*.
2. **Hardware Setting**

DFR0023 (LM35) is used here to measure the environmental temperature. The sensor has three pins: VCC, Output, and GND. The VCC should be connected to 5 V and the GND to a common ground with your Arduino. The signal output of DFR0023 should be plugged into a defined pin on the DFRobot UNO R3 board (here, Analog 0) (Fig. 3.9).

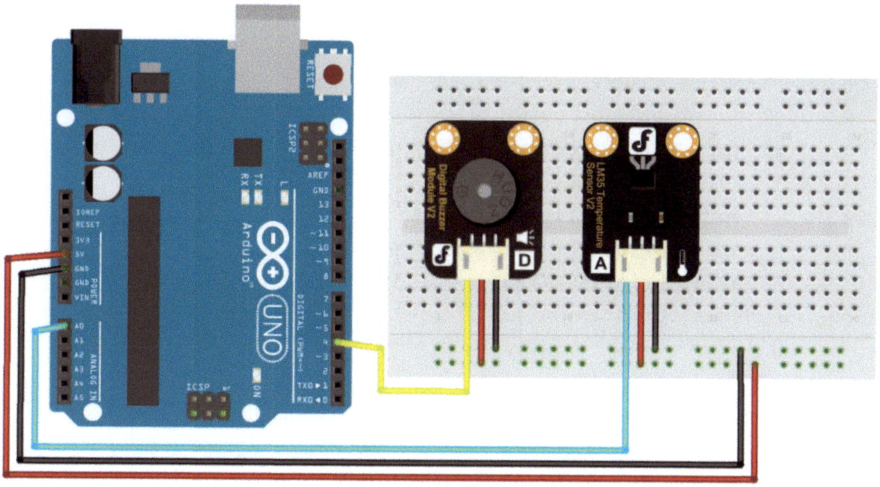

Fig. 3.9 A diagram of the layout of the DFR0023, DFR0032, and UNO R3

3. Sample Codes

```
int ALARM = 4; //define Buzzer digital pin 4
int LM35 = 0; // connect LM35 to the analog input pin 0
float tmprVal = 0; //define value
float sinVal;
int toneVal;
unsigned long tepTimer;

void setup(void) {
  Serial.begin(9600); // configure baud rate to 9600 bps
  pinMode(ALARM, OUTPUT); // configure pin of buzzer
}
void loop(void) {
  int val = analogRead(LM35); // read LM35
  tmprVal = (float) val * (5 / 10.24); //Convert the voltage value
to temperature value
  if (tmprVal > 27) { //If temperature > 27, the buzzer starts to
make sound.
    for (int x = 0; x < 180; x++) {
      //Convert sin function to radian
      sinVal = (sin(x * (3.1412 / 180)));
      //Use sin function to generate frequency of sound
      toneVal = 2000 + (int(sinVal * 1000));
      //Configure the buzzer pin 4
      tone(ALARM, toneVal);
      delay(2);
    }
  }
  else { // If the temperature <= 27, turn off the buzzer
    noTone(ALARM); // Turn off the buzzer
  }
  if (millis() - tepTimer > 50) { // Every 500 ms, serial port
outputs temperature value.
    tepTimer = millis();
    Serial.print("temperature: ");
    Serial.print(tmprVal);
    Serial.println("C");
  }
}
```

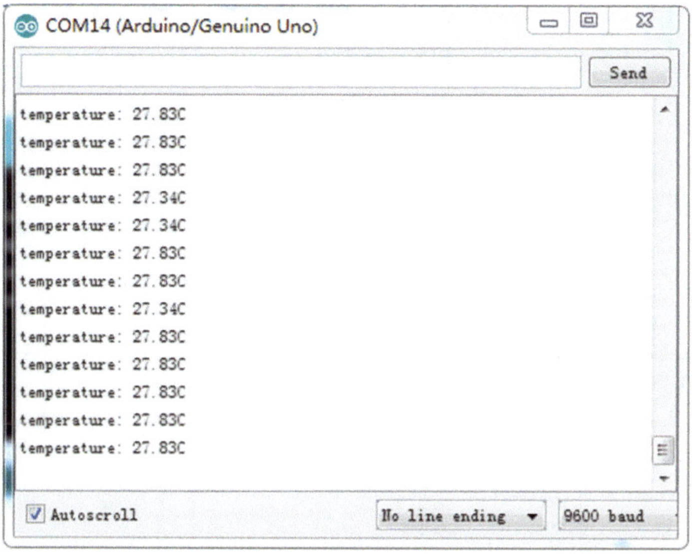

Fig. 3.10 Log messages from Arduino using DFR0023

4. **Results**

After the code is successfully uploaded, the serial monitor of the Arduino IDE should exhibit values as follows (Fig. 3.10).

Once the temperature reaches 27 °C, the buzzer starts to sound. If the temperature drops below 27 °C, the buzzer stops.

3.4 Temperature and Humidity Sensor

3.4.1 Introduction

DHT11 is a four PIN sensor (Fig. 3.11) that can measure temperatures ranging from 0 to 50 °C and relative humidity ranging from 20 to 95%. The sensor uses its own proprietary 1-Wire protocol to communicate with Arduino and works at 3.3–5 V.

The timings must be precise and according to the datasheet of the sensor. Each DHT11 sensor is strictly calibrated in a laboratory to ensure absolute accuracy for humidity calibration. The calibration coefficients are stored as programmers in the OTP memory, which are used by the sensor's internal signal detection process. The single-wire serial interface makes system integration rapid and easy. Its small size, low power consumption, and up-to-20 m signal transmission range makes it the best choice for various applications, including highly demanding ones.

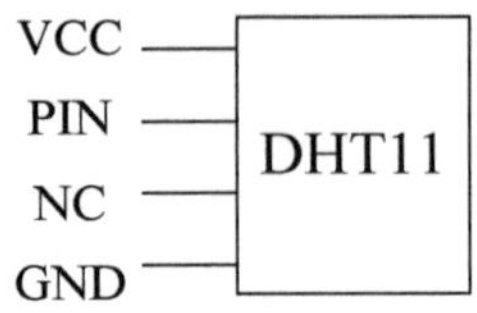

Fig. 3.11 Block diagram of DHT11

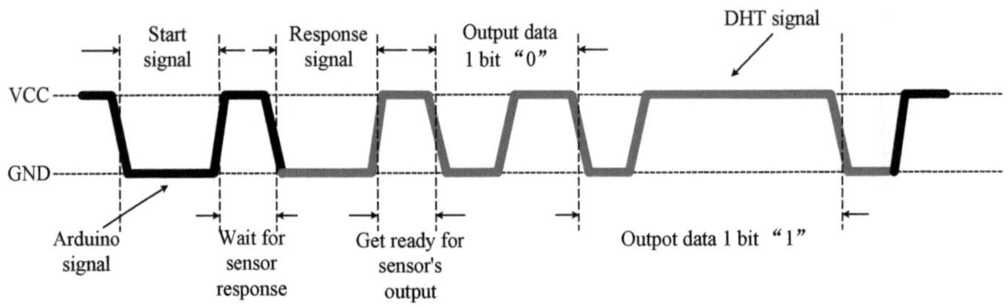

Fig. 3.12 Overall communication process

Figure 3.12 shows an overall communication process between Arduino and DHT11. When Arduino sends a start signal, DHT11 changes from the low power consumption mode to the working mode, waiting for Arduino to complete the start signal. Once it is complete, DHT11 sends a response signal of 40-bit data that includes relative humidity and temperature information to Arduino.

Arduino initiates the data transmission process by pulling the data bus low for about 18 ms and keeps it High for about 20–40 μs before releasing it. Subsequently, the sensor responds to the Arduino's data transfer request by pulling the data bus Low for 80 μs followed by 80 μs of High. At this point, Arduino is ready to receive data from the sensor (Fig. 3.13).

The format of 40-bit data is "8 bit humidity integer data + 8 bit Humidity decimal data + 8 bit temperature integer data + 8 bit fractional temperature data + 8 bit parity bit." The received data is correct when the 8 bit checksum is equal to the results of the last eight. The example is as follows:

00110101	00000000	00011000	00000000	01001101
High humidity	Low humidity	High temperature	Low temperature	Pairy bit

Calculate:

00110101 + 00000000 + 00011000 + 00000000 = 01001101

Received data is correct, and

Humidity: $00110101 = 0 \times 35 = 53\%RH$

Temperature: $00011000 = 0 \times 18 = 24\ °C$

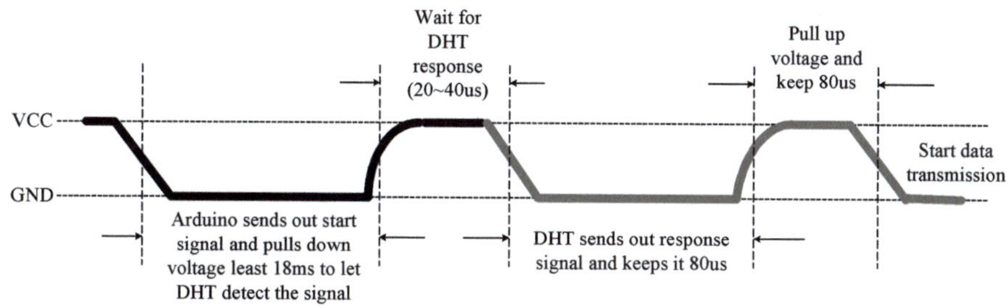

Fig. 3.13 Send start signal to DHT

3.4.2 *Demonstration*

1. **Components**
   - DFRobot UNO R3 board and USB cable × 1.
   - DFR0067 (DHT11 Temperature and Humidity Sensor V2) × 1.
   - Jumper wires × *n*.
2. **Hardware Setting** (Fig. 3.14)

Fig. 3.14 A diagram of the layout of the DFR0067 and UNO R3

3. **Sample Codes**

```
#include <dht11.h>
#define DHT11_PIN 4 //define humidity digital pin 4
dht11 DHT;
void setup() {
  Serial.begin(9600); // configure baud rate to 9600 bps
  Serial.println("DHT TEST PROGRAM ");
  Serial.print("LIBRARY VERSION: ");
  Serial.println(DHT11LIB_VERSION);
  Serial.println();
  Serial.println("Type,\tstatus,\tHumidity (%),\tTemperature (C)");
}
void loop() {
  int chk;
  Serial.print("DHT11, \t");
  chk = DHT.read(DHT11_PIN); // read data
  switch (chk) {
    case DHTLIB_OK:
      Serial.print("OK,\t");
      break;
    case DHTLIB_ERROR_CHECKSUM:
      Serial.print("Checksum error,\t");
      break;
    case DHTLIB_ERROR_TIMEOUT:
      Serial.print("Time out error,\t");
      break;
    default:
      Serial.print("Unknown error,\t");
      break;
  }
  Serial.print(DHT.humidity, 1); // display data
  Serial.print(",\t");
  Serial.println(DHT.temperature, 1);
  delay(1000);
}
```

Before we verify these codes with Arduino, the DHT11Lib, i.e., "dht11.h", should be downloaded first (http://playground.arduino.cc/Main/DHTLib). The uploading path is "Sketch → Include Library → Add.Zip Library."

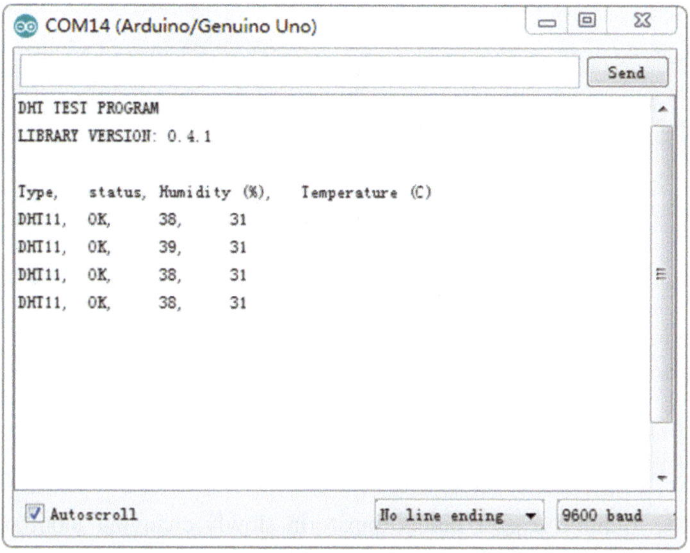

Fig. 3.15 Log messages from Arduino using DFR0067

4. **Results**

After uploading the code, you can see the humidity and temperature values through the serial monitor (Baudrate = 9600) (Fig. 3.15).

3.5 Line-Tracking Sensor

3.5.1 Introduction

Line tracking is the most basic function of a smart mobile robot. The line-tracking sensor can detect white lines in black and black lines in white via the TTL signal (Black for LOW output, White for HIGH output). Line-tracking sensor SEN0017 is used here to explain the tracking principle, whose schematic is shown in Fig. 3.16.

ITR20001/T consists of an infrared emitting diode and an NPN silicon phototransistor, encased side-by-side on a converging optical axis in a black thermoplastic housing. The phototransistor receives radiation from the IR only. This is in normal conditions. However, when a reflecting object is in close proximity with the ITR, the phototransistor receives reflected radiation. 74HCT14 is a hex inverter with Schmitt-trigger inputs. This device features reduced input threshold levels to allow interfacing to TTL logic levels. Inputs also include clamp diodes, which enables the use of current limiting resistors to interface inputs to voltages in excess

Fig. 3.16 Schematic of SEN0017

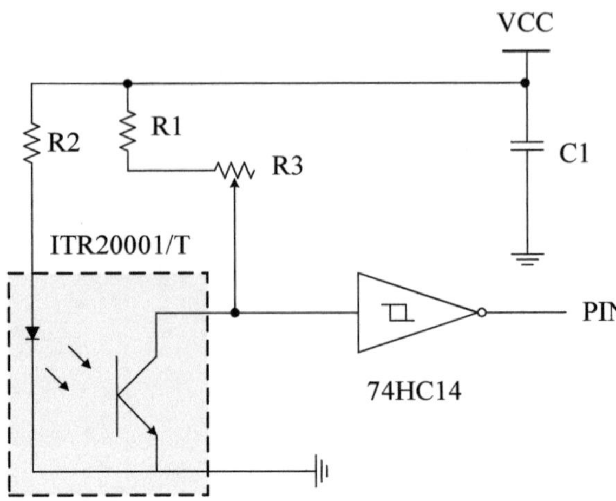

of the VCC. Schmitt-trigger inputs transform slowly changing input signals into sharply defined jitter-free output signals. The best distance between objects such as the ground and sensor is 1–2 cm. For the purpose of tracking lines in different cases, an optional multi-channel mix can be implemented with the necessary line tracking sensors. The IR is absorbed by black lines and reflected by white lines, which is why we have Black for LOW output and White for HIGH output.

3.5.2 Demonstration

1. **Components**

   - DFRobot UNO R3 board and USB cable × 1.
   - SEN0017 (line-tracking sensor) × 1.
   - Jumper wires × *n*.

2. **Hardware Setting**

The line-tracking sensor SEN0017 has three pins: VCC, Output, and GND. The VCC should be connected to 5 V and the GND to a common ground with your Arduino. The Output pin of SEN0017 should be plugged into a defined pin on the DFRobot UNO R3 board (here, PIN 2) (Fig. 3.17).

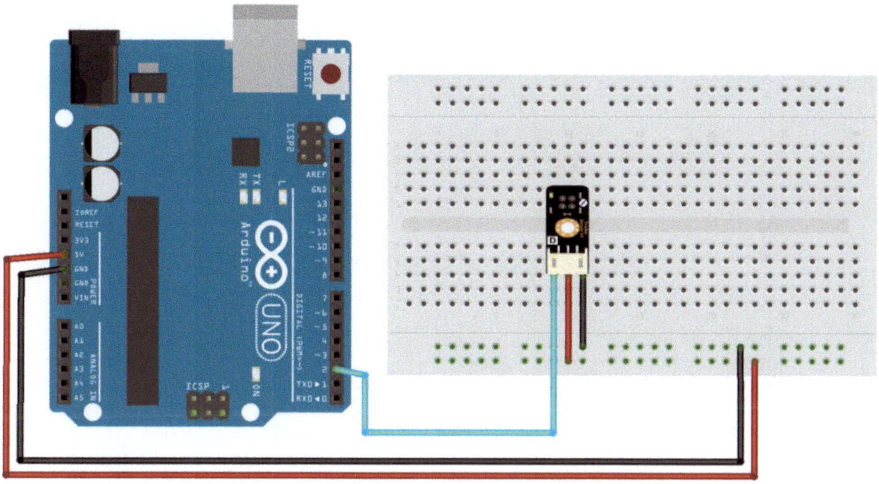

Fig. 3.17 A diagram of the layout of the SEN0017 and UNO R3

3. **Sample Codes**

```
#define LINE 2 //define line tracking digital pin 2
int val = 0;
void setup() {
  Serial.begin(9600); // configure baud rate to 9600 bps
}
void loop() {
  val = digitalRead(LINE);
  Serial.println(val);
  delay(500);
}
```

4. **Results**

It should be noted that the best distance between objects such as ground and the sensor is 1–2 cm. When the sensor detects a black line, then the light in the SEN0017 board is off and the Output will be LOW (see the serial monitor). Else, the light in the SEN0017 board is on and the Output is HIGH (see the serial monitor) (Fig. 3.18).

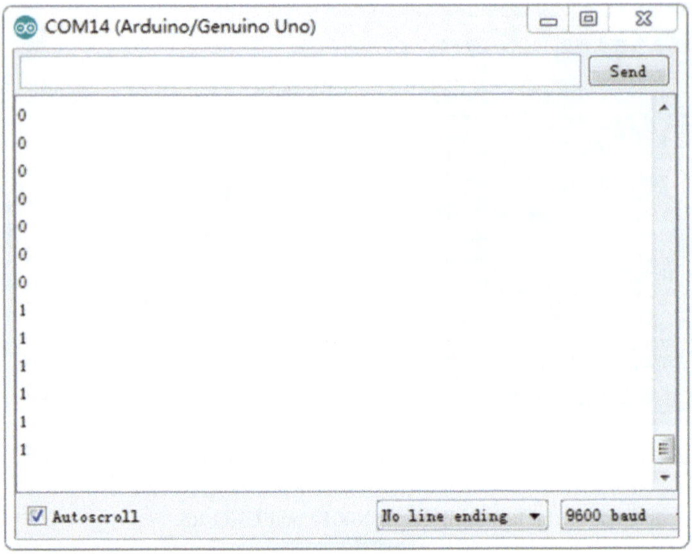

Fig. 3.18 Log messages from Arduino using SEN0017

3.6 Ultrasonic Sensors

3.6.1 Introduction

When creating an autonomous target tracking robot, one of the most crucial aspects is obstacle detection and avoidance. Often, a target may move in a scenario where there is an object between the target and robot. A sensor must be able to detect the object, with adequate range to allow the robot to respond and move accordingly. Ideally, the sensor must be small, low in cost, and easy to manufacture and must be usable on a large scale. A readily available sensor that fits all of these requirements is the ultrasonic sensor.

The ultrasonic sensor uses sonar to determine the distance from an object, similar to bats and dolphins. The sensor is a non-contact distance measurement equipment that uses a timing variable *t*, measured from the transmission of some kind of a pilot signal till a reflected signal is received from the object (Fig. 3.19).

The distance *L* to the object can be calculated as follows:

$$L = \frac{vt\cos\theta}{2} \tag{3.3}$$

where, v is the speed of ultrasonic waves in the media (i.e., 340 m/s), θ is the reflection angle. If the transmitter and the receiver are close to each other, then $\cos\theta \approx 1$.

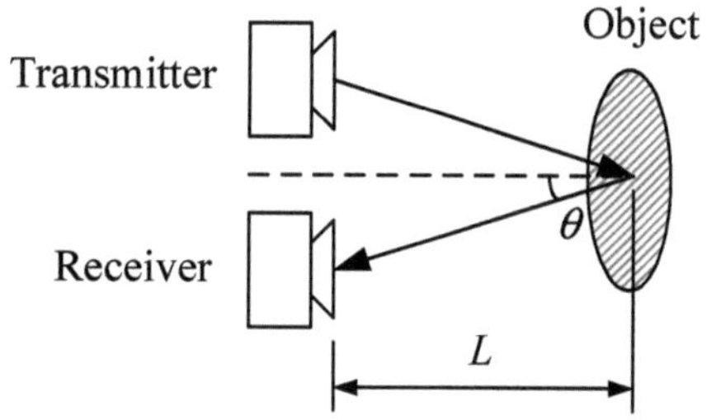

Fig. 3.19 The principle of ultrasonic sensor

For example, if the object is L = 10 cm away from the sensor, and the speed of sound is 340 m/s or 0.034 cm/μs, the sound wave will need to travel about 294 μs. However, what you will get from the receiver will be double that number because the sound wave needs to travel forward and bounce backward. Therefore, in order to obtain the distance in cm we need to multiply the received travel time value from the receiver by 0.034 cm/μs and divide it by 2.

3.6.2 HC-SR04

Ultrasonic Sensor HC-SR04 is used here to measure distances in the range of 2 m–400 cm with an accuracy of 3 mm. The sensor module consists of an ultrasonic transmitter, receiver, and control circuit. The working principle of an ultrasonic sensor is as follows:

1. A high level signal is sent for 10 μs using the Trigger.
2. The module sends eight 40 kHz signals automatically, and then detects whether the pulse is received or not.
3. If the signal is received, then it forms a high level. The time of the high duration is the time gap between sending and receiving the signal.
4. The distance is calculated using Eq. (3.3).

The timing diagram is shown as Fig. 3.20.

3.6.3 Demonstration

1. **Components**

- DFRobot UNO R3 board and USB cable × 1.
- HC-SR04 (ultrasonic distance sensor) × 1.
- Jumper wires × n.

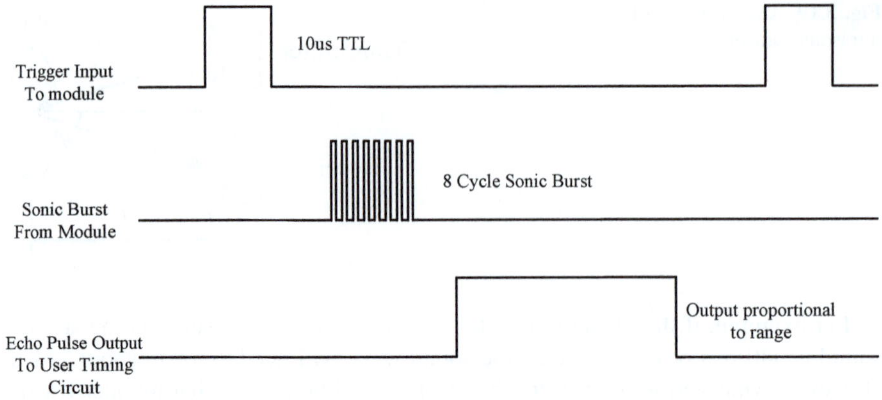

Fig. 3.20 Timing diagram

2. **Hardware Setting**

The ultrasonic sensor HC-SR04 has four pins: VCC, TRIG, ECHO, and GND. The VCC should be connected to 5 V and the GND to a common ground with your Arduino. TRIG is used for emitting the transmitted signal. The output pin of the Ultrasonic sensor should be plugged into a defined pin on the DFRobot UNO R3 board. The ECHO pin is what is used for detecting the response from the ultrasonic wave; this two should be plugged into a separate defined pin on the DFRobot UNO R3 board (Fig. 3.21).

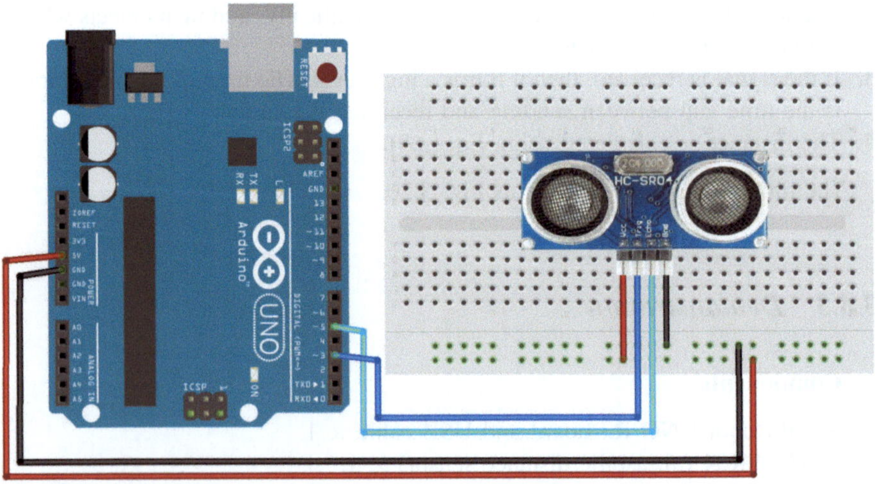

Fig. 3.21 A diagram of the layout of the HC-SR04 and UNO R3

3. Sample Codes

```
const int trigPin = 5; // PWM trigger
const int echoPin = 3; // PWM Output 0-25000US,Every 50US represent 1cm
long duration;       // defines variables
int distance;
void setup() {
  pinMode(trigPin, OUTPUT); //Sets the trigPin as an Output
  pinMode(echoPin, INPUT); // Sets the echoPin as an Input
  Serial.begin(9600); //configure baud rate to 9600 bps
}
void loop() {
  digitalWrite(trigPin, LOW); //Clears the trigPin
  delayMicroseconds(2);
  // Sets the trigPin on HIGH state for 10 micro seconds
  digitalWrite(trigPin, HIGH);
  delayMicroseconds(10);
  digitalWrite(trigPin, LOW);
  duration = pulseIn(echoPin, HIGH); // Reads the echoPin, returns the sound wave travel time in microseconds
  distance = duration * 0.034 / 2; // Calculating the distance

  Serial.print("Distance Measured="); // Prints the distance on the Serial Monitor
  Serial.print(distance);
  Serial.println("cm");
}
```

4. Results

Open the IDE serial port (Baudrate = 9600); the distance is displayed on it (Fig. 3.22).

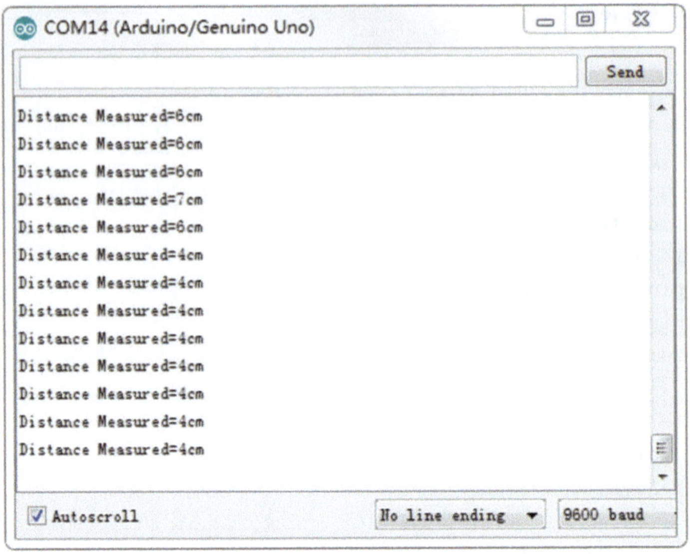

Fig. 3.22 Log messages from Arduino using HC-SR04

3.7 Digital Infrared Motion Sensor

3.7.1 Introduction

A motion sensor is an electronic device that is used for motion detection. It contains an electronic sensor that quantifies motion and can either be integrated with or connected to other devices that alert a user of the presence of a moving object that is within the field of view of the sensor. Motion sensors are a vital component of comprehensive security systems for businesses and homes. An electronic motion detector contains a motion sensor that transforms detected motion into an electric signal. Infrared motion sensors allow you to sense motion; it almost always is used to detect whether a human has moved in or out of a sensor's range. They are small, inexpensive, low power, easy to use, and do not wear out. For that reason they are commonly found in appliances and gadgets used in homes or businesses.

3.7.2 Demonstration

1. **Components**

   - DFRobot UNO R3 board and USB cable × 1.
   - SEN0018 (digital infrared motion sensor) × 1.
   - LED × 1.

- Resistor (220 Ω) × 1.
- Jumper wires × *n*.

2. **Hardware Setting**

The digital infrared motion sensor SEN0018 has three pins: VCC, Pinout, and GND. The VCC should be connected to 5 V and the GND to a common ground with your Arduino. The Pinout is used for emitting the transmitted signal.

Power SEN0018 up and wait 12 s for the sensor to obtain a snapshot of the still room. If anything moves after that period, the Pinout will go low (Fig. 3.23).

3. **Sample Codes**

```
const int InfraredPin = 2; //define Infrared sensor pin 2
const int LED = 13; //define LED digital pin 13
void setup() {
  pinMode(LED, OUTPUT);
  pinMode(InfraredPin, INPUT);
  Serial.begin(9600); //configure baud rate to 9600 bps
}
void loop() {
  if (digitalRead(InfraredPin) == HIGH) {
    digitalWrite(LED, HIGH);
    Serial.println("anything move in!");
  }
  else {
    digitalWrite(LED, LOW);
    Serial.println("this is nothing!");
  }
  delay(1000);
}
```

4. **Results**

When the sensors detect that people have moved, the light illuminates (Fig. 3.24).

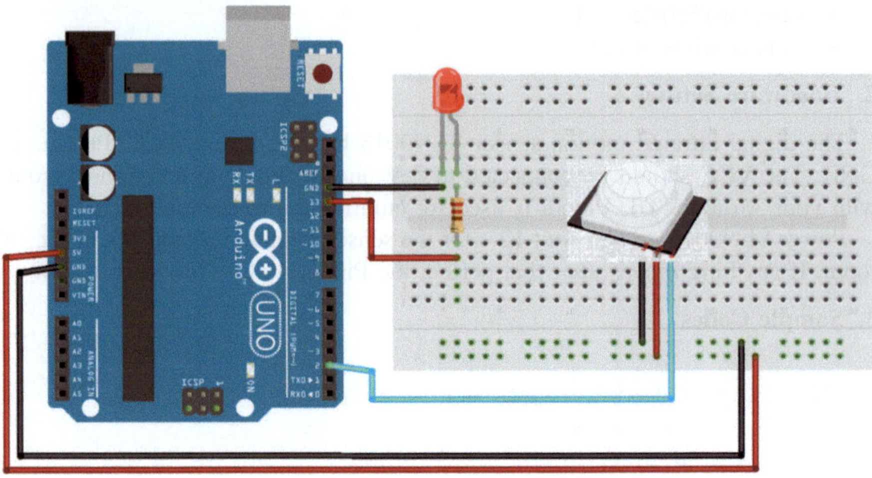

Fig. 3.23 A diagram of the layout of the SEN0018 and UNO R3

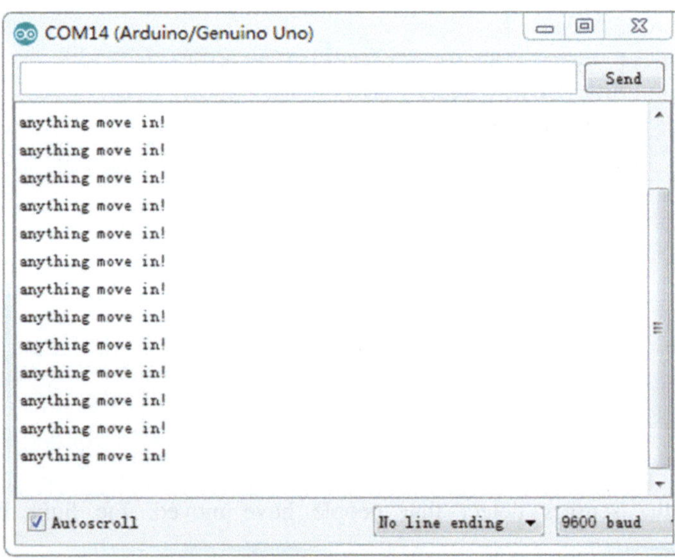

Fig. 3.24 Log messages from Arduino using SEN0018

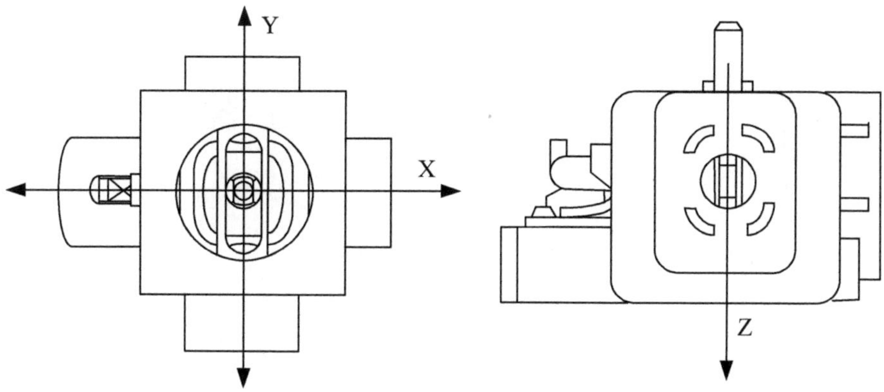

Fig. 3.25 The diagrams of *X*, *Y*, and *Z* axes of Joystick

3.8 Joystick Module

3.8.1 Introduction

A number of robots contain a joystick module. It simply connects to two analog outputs based on your commands with X and Y control. The joystick provides an affordable solution to move robots to a target directly. It also has a digital output pin with 0, 1 (*Z* axis).

The joystick module is made up of two potentiometers and a press button, which are associated with the *X* and *Y* axes, respectively. The output voltages are changed between 0 and 5 V when the potentiometer is moved in the direction of X or Y. The digital output is changed to 0 when the button is pressed. The diagrams of a joystick module are shown in Fig. 3.25.

3.8.2 Demonstration

1. **Components**

   - DFRobot UNO R3 board and USB cable × 1.
   - DFR0061 (joystick module) × 1.
   - Jumper wires × *n*.

2. **Hardware Setting**

The joystick module has nine pins (three pins for each axis): VCC, Output, and GND. In the *X* and *Y* axes, the VCC should be connected to 5 V and the GND should be connected to a common ground on the DFRobot UNO R3 board. The Output pin

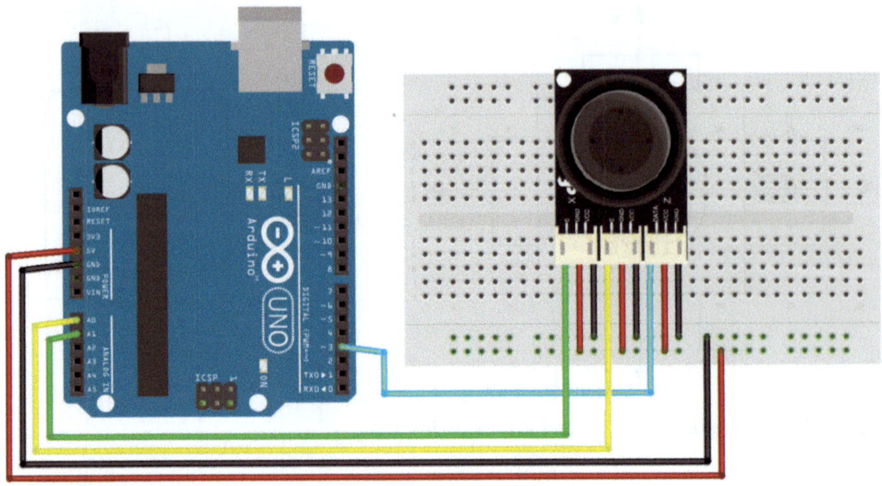

Fig. 3.26 A diagram of the layout of the DFR0061 and UNO R3

of the *X*/*Y* axis should be plugged into the analog pin (here, PIN 0/1) and the Output of the *Z* axis should be connected to the digital pin (here, PIN 3) (Fig. 3.26).

3. **Sample Codes**

```
int JoyStick_X = 1; //define X-Axis analog pin 1
int JoyStick_Y = 0; //define Y-Axis analog pin 0
int JoyStick_Z = 3; //define X-Axis digital pin 3
void setup() {
  pinMode(JoyStick_Z, INPUT);
  Serial.begin(9600);//configure baud rate to 9600 bps
}
void loop() {
  int x, y, z;
  x = analogRead(JoyStick_X);
  y = analogRead(JoyStick_Y);
  z = digitalRead(JoyStick_Z);
  Serial.print(x , DEC);
  Serial.print(",");
  Serial.print(y , DEC);
  Serial.print(",");
  Serial.println(z , DEC);
  delay(100); // wait for 100 ms
}
```

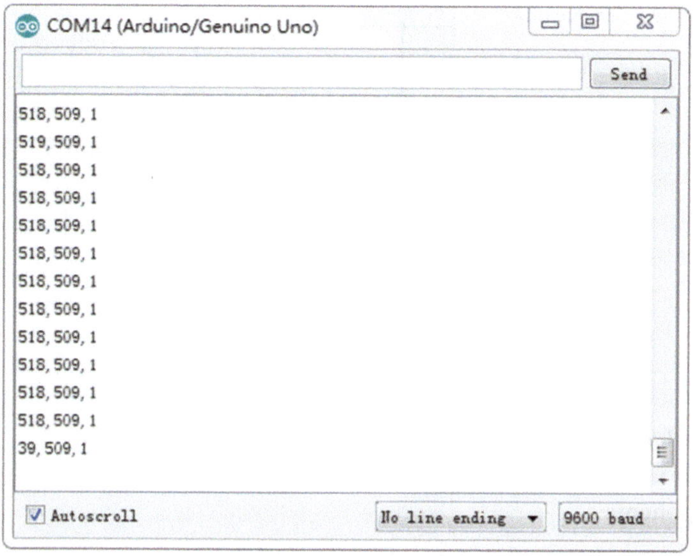

Fig. 3.27 Log messages from Arduino using DFR0061

4. **Results**

After uploading the program to your Arduino IDE, opening the serial monitor (Baudrate = 9600), and moving the joystick, you can observe the following.

In the Arduino sketch, the analogRead () function returns a number in the 0–1023 range (512 at the center/idle position of the joystick) (Fig. 3.27).

3.9 Gas Sensor

3.9.1 Introduction

Gas sensors are wieldy used to detect gas leakages of LPG, i-Butane, Propane, Methane, Alcohol, Hydrogen, and smoke in houses and factories. The gas sensor is composed of a micro AL_2O_3 ceramic tube, Tin Dioxide sensitive layer, electrode and heater measurer, which are fixed into a sensitive component. The conductivity of the gas sensing layer changes when a gas leakage occurs, which changes the voltage value in the output. The heater provides the necessary work condition for sensitive components. The schematic is shown as follow (Fig. 3.28).

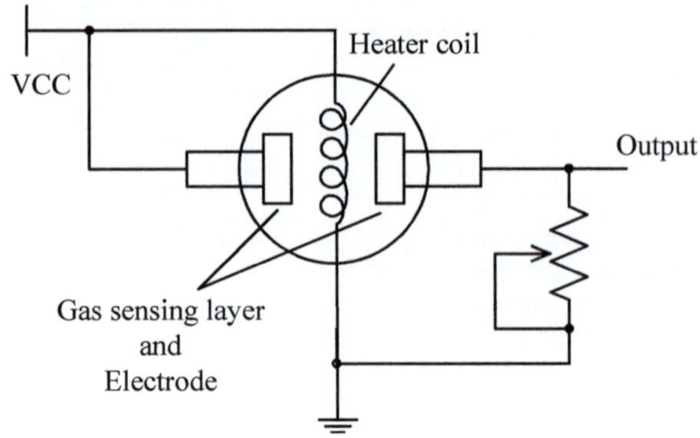

Fig. 3.28 The schematic of a gas sensor

3.9.2 *Demonstration*

1. **Components**

- DFRobot UNO R3 board and USB cable × 1.
- DFR0049 (gas sensor) × 1.
- Jumper wires × *n*.

2. **Hardware Setting**

The gas sensor DFR0049 uses three pins to connect with Arduino: VCC, Output, and GND. The VCC should be connected to 5 V and the GND should be plugged into the common ground. The Output pin should be connected to the analog pin (here, A0) (Fig. 3.29).

3. **Sample Codes**

```
int GasPin = 0; //define gas sensor analog pin 0
int val; //define value
void setup() {
  Serial.begin(9600);//configure baud rate to 9600 bps
}
void loop() {
  val = analogRead(GasPin);  //Read Gas value
  Serial.println(val, DEC);
  delay(100); // wait for 100 ms
}
```

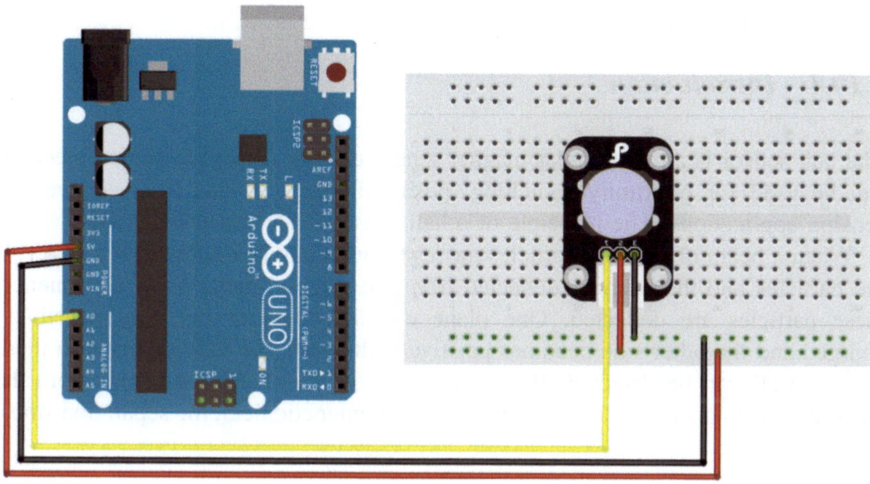

Fig. 3.29 A diagram of the layout of the DFR0049 and UNO R3

4. **Results**

After uploading the program to your Arduino IDE, opening the serial monitor (Baudrate = 9600), and generating some smoke, you should see the following values (Fig. 3.30).

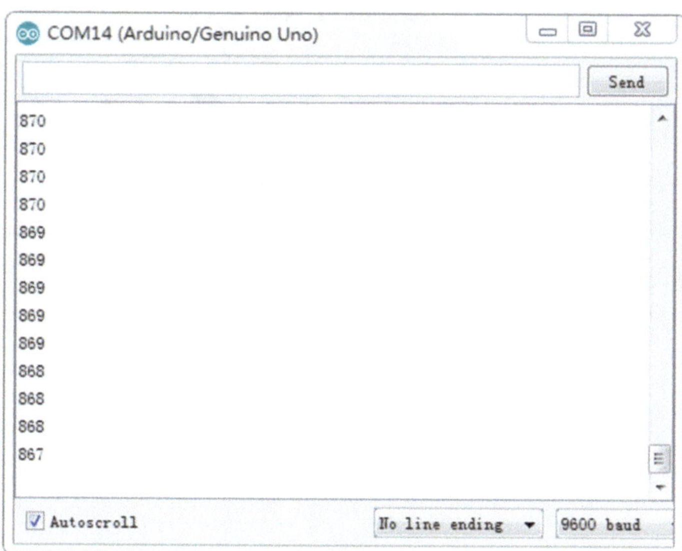

Fig. 3.30 Log messages from Arduino using DFR0049

3.10 Hall Sensor

3.10.1 Introduction

The hall sensor is an Omnipolar magnet sensor used to detect magnetic objects. It can be used for proximity switching, positioning, speed detection, and current sensing applications. The principle of a hall sensor is shown in Fig. 3.31.

When a beam of charged particles passes through a magnetic field, forces act on the particles and the beam is deflected from the original path. As a consequence, these particles are deflected. One plane of the conductor becomes negatively charged and the other side becomes positively charged. The voltage between these planes (VH) can be detected. When the force on the charged particles from the electric field balances the force produced by the magnetic field, the separation stops.

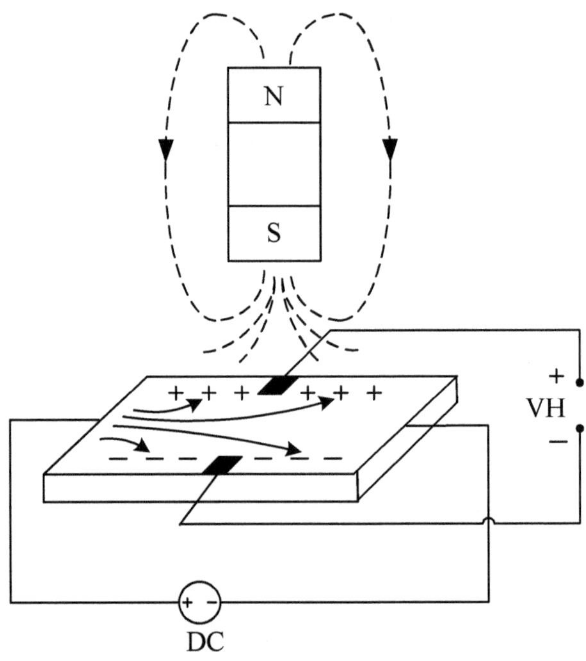

Fig. 3.31 The working principle of a hall sensor

3.10.2 *Demonstration*

1. **Components**

- DFRobot UNO R3 board and USB cable × 1.
- SEN0185 (hall sensor) × 1.
- LED × 1.
- Resistor (220 Ω) × 1.
- Jumper wires × *n*.

2. **Hardware Setting**

SEN0185 uses three pins to connect with Arduino: VCC, Output, and GND. The VCC should be connected to 5 V and the GND should be plugged into the common ground. The Output pin should be connected to the digital pin (here, PIN2) (Fig. 3.32).

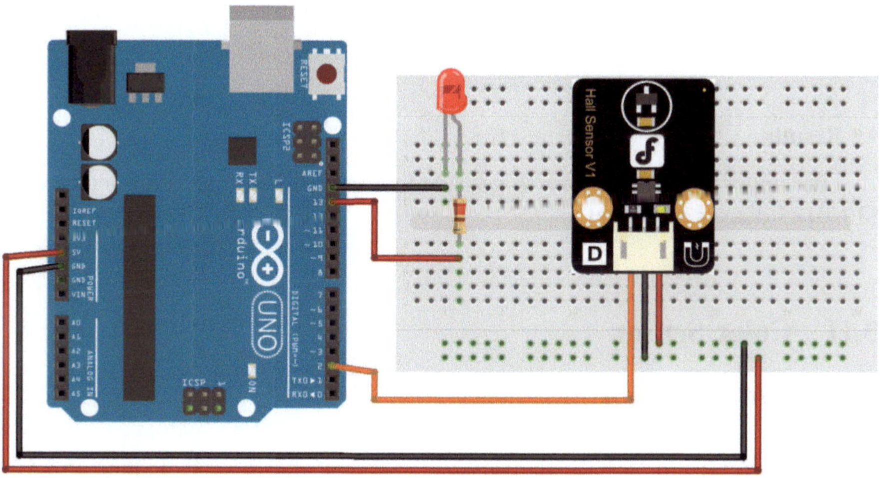

Fig. 3.32 A diagram of the layout of the SEN0185 and UNO R3

3. **Sample Codes**

```
int ledPin = 13; // choose the pin for the LED
int hallPin = 2; // choose the input pin
int val = 0; // variable for reading the pin status
void setup() {
  pinMode(ledPin, OUTPUT); // declare LED as output
  pinMode(hallPin, INPUT); // declare pushbutton as input
}
void loop() {
  val = digitalRead(hallPin); // read input value
  if (val == HIGH) { // check if the input is HIGH
    digitalWrite(ledPin, HIGH); // turn LED ON
  }
  else {
    digitalWrite(ledPin, LOW); // turn LED OFF
  }
}
```

4. **Results**

After uploading the program to your Arduino, the digital light illuminates when SEN0185 detects a magnet.

3.11 Color Sensor

3.11.1 Introduction

The TCS3200 Color Sensor is a complete color detector, including a TAOS TCS3200 RGB sensor chip and four white LEDs. It can detect and measure a nearly limitless range of visible colors. The applications of TCS3200 include test strip reading, sorting by color, ambient light sensing and calibration, and color matching, etc.

In TCS3200, the light-to-frequency converter reads an 8×8 array of photodiodes. 16 photodiodes have blue filters, 16 photodiodes have green filters, 16 photodiodes have red filters, and 16 photodiodes are clear with no filters (shown in Fig. 3.33). The four types (colors) of photodiodes are interdigitated to minimize the effect of non-uniformity of incident irradiance. All photodiodes of the same color are connected in parallel. When choosing a color filter, only one particular color is permitted through and the other colors are prevented. The color that needs to be sensed by a color sensor is selected by two pins S2 and S3 (shown in Table 3.1).

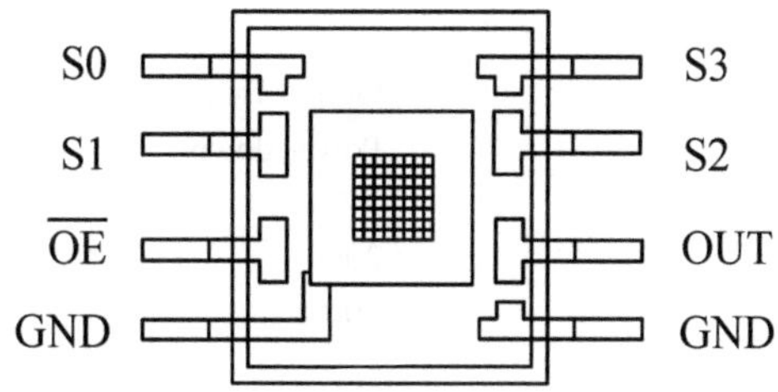

Fig. 3.33 The TCS3200 chip

Table 3.1 Selectable options

S2	S3	Photodiode type
L	L	Red
L	H	Blue
H	L	Clear (no filter)
H	H	Green

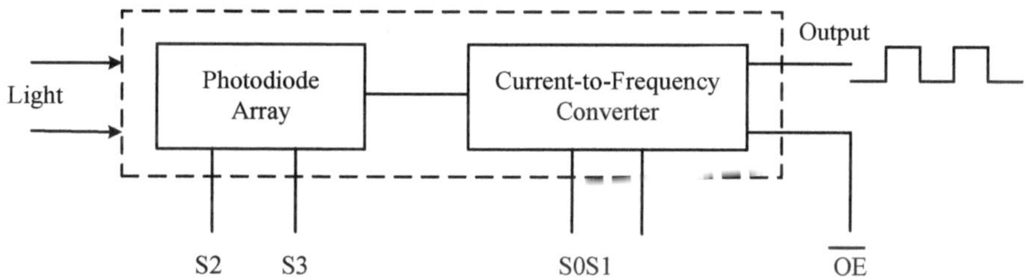

Fig. 3.34 The control system inside the TCS300 module

Then, signals from the sensor can calculate the RGB values in terms of the red, blue, and green components.

If we need to sense the RED color intensity, we need to set both pins to LOW. Once that is done the sensor detects the intensity and sends the value to the control system inside the module.

The control system inside the module is shown in Fig. 3.34. The light intensity measured by the array is sent as a current for frequency conversion. What it does is, it generates a square wave whose frequency relates to the current sent by ARRAY.

Therefore, we have a system that generates a square wave whose frequency depends on the light intensity of the color that is selected by S2 and S3.

The signal frequency sent by the module can be modulated depending on the use. We can change the output signal frequency bandwidth (Table 3.2).

Frequency scaling is performed by two bits S0 and S1. For convenience, we limit the frequency scaling to 20%. This is done by setting S0 to high and S1 to LOW. This feature comes is useful when we use the module on a system with a low clock.

Table 3.2 The signal frequency selection

S0	S1	Output frequency scaling (f0)
L	L	Power down
L	H	2%
H	L	20%
H	H	100%

3.11.2 Demonstration

1. **Components**

- DFRobot UNO R3 board and USB cable × 1.
- SEN0101 (TCS3200 color sensor) × 1.
- Jumper wires × *n*.

2. **Hardware Setting**

The TCS3200 sensor contains 10 pins, 7 of them are used here: S0, S1, S2, S3, Output, VCC, and GND. S0, S1, S2, S3, and Output should be connected to the digital pin (here: PIN 3, 4, 5, 6, and 2, respectively), the VCC should be connected to 5 V, and the GND should be plugged into the ground (Fig. 3.35).

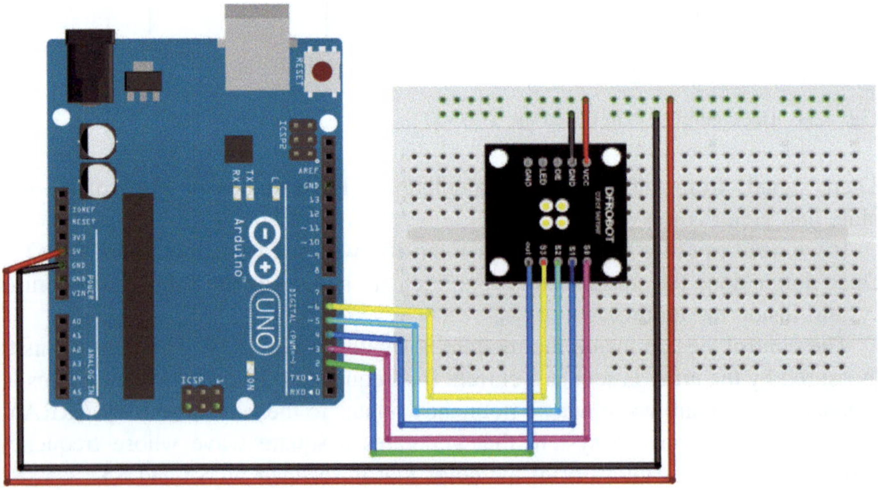

Fig. 3.35 A diagram of the layout of the SEN0101 and UNO R3

3. **Sample Codes**

```
int s0 = 3, s1 = 4, s2 = 5, s3 = 6;
int OutPut = 2; //define pin 2 of as output
unsigned int frequency = 0;
void setup() {
  pinMode(s0, OUTPUT); //PINS 3,4,5,6 as OUTPUT
  pinMode(s1, OUTPUT);
  pinMode(s2, OUTPUT);
  pinMode(s3, OUTPUT);
  digitalWrite(s0, HIGH); //setting frequency as 20%
  digitalWrite(s1, LOW);
  Serial.begin(9600);//configure baud rate to 9600 bps
}
void loop() {
  digitalWrite(s2, LOW); //setting for RED color sensor
  digitalWrite(s3, LOW);
  frequency = pulseIn(OutPut, LOW); //reading frequency
  Serial.print("Red=");
  Serial.println(frequency, DEC);
  delay(500);
  digitalWrite(s2, LOW); //setting for BLUE color sensor
  digitalWrite(s3, HIGH);
  frequency = pulseIn(OutPut, LOW); //reading frequency
  Serial.print("Blue=");
  Serial.println(frequency, DEC);
  delay(500);
  digitalWrite(s2, HIGH); //setting for GREED color sensor
  digitalWrite(s3, HIGH);
  frequency = pulseIn(OutPut, LOW); //reading frequency
  Serial.print("Green=");
  Serial.println(frequency, DEC);
  delay(500);
}
```

4. **Results**

After uploading the program to your Arduino, placing some colored objects close to the sensor, and opening the serial monitor (Baudrate = 9600), three color intensities are shown as follows (Fig. 3.36).

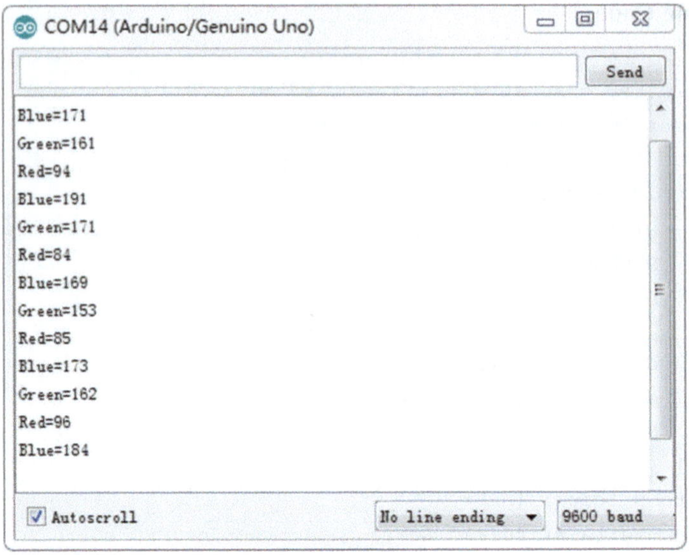

Fig. 3.36 Log messages from Arduino using SEN0101

3.12 Digital Tilt Sensor

3.12.1 Introduction

The digital tilt sensor is a digital tilt switch. There are two modules of the tilt sensor, mercury-based and ball-based. Because mercury is a toxic substance, we only consider the ball-based sensor as an example. The tilted sensor makes the ball move to the switch on account of gravity and makes the switch close (Fig. 3.37).

It should be noticed that the sensor is simply a simple digital tilt sensor, which outputs '0' or '1'. The tilt angle could not be adjusted. Moreover, strong vibrations can affect its output.

3.12.2 Demonstration

1. **Components**

- DFRobot UNO R3 board and USB cable × 1.
- DFR0028 (digital tilt sensor) × 1.
- LED × 1.
- Resistor (220 Ω) × 1.
- Jumper wires × *n*.

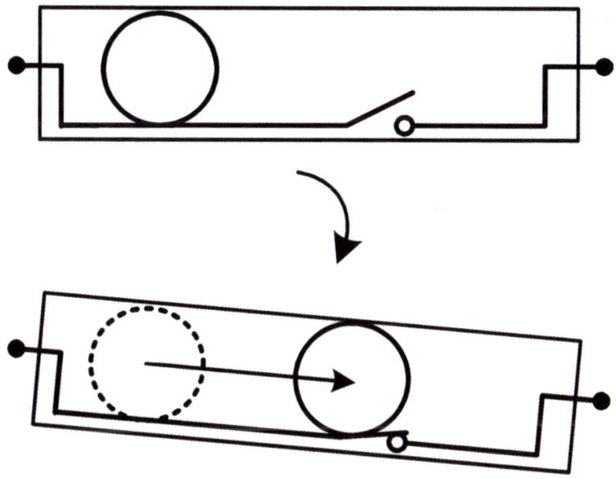

Fig. 3.37 The working principle of a tilt sensor

2. **Hardware Setting**

The tilt sensor has three pins: VCC, Output, and GND. The VCC should be connected to 5 V and the GND should be plugged into the common ground. The Output pin should be connected to the digital pin (here, PIN3) (Fig. 3.38).

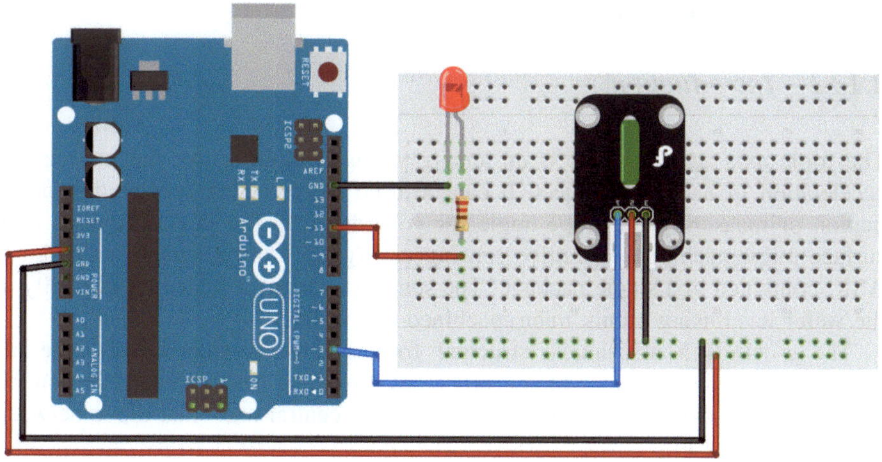

Fig. 3.38 A diagram of the layout of the DFR0028 and UNO R3

3. **Sample Codes**

```
int ledPin = 13; // Connect LED to pin 13
int switcher = 3; // Connect Tilt sensor to Pin3
void setup() {
  pinMode(ledPin, OUTPUT); // Set pin 13 to output mode
  pinMode(switcher, INPUT); // Set pin 3 to input mode
}
void loop() {
  if (digitalRead(switcher) == HIGH) //Read sensor value
    digitalWrite(ledPin, HIGH); // Turn on LED when the sensor is
tilted
  else
    digitalWrite(ledPin, LOW); // Turn off LED when the sensor is
not triggered
}
```

4. **Results**

After uploading the program to your Arduino, the digital light illuminates when DFR0028 is tilted to the switch side.

3.13 Triple Axis Acceleration Sensor

3.13.1 Introduction

The triple axis accelerometer is an analog voltage output sensor that detects the acceleration of a moving object. It can be used for detecting acceleration information or measuring the slope angle for your devices. The device consists of a surface micromachined capacitive sensing cell (g-cell) and a signal conditioning ASIC contained in a single package. The sensing element is sealed hermetically at the wafer level using a bulk micromachined cap wafer.

The g-cell is a mechanical structure formed from semiconductor materials (polysilicon) using semiconductor processes (masking and etching). It can be modeled as a set of beams attached to a movable central mass that move between fixed beams (see Fig. 3.39).

As the center beam moves with acceleration, the distance between the beams changes and each capacitor's value changes. It can be calculated by:

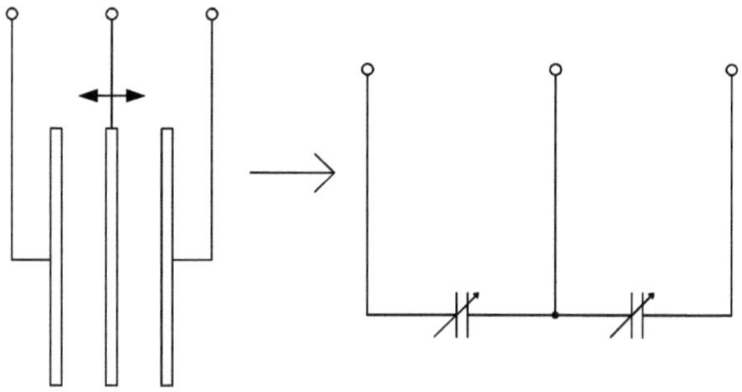

Fig. 3.39 The simplified model of a g-cell structure

Table 3.3 g-select pin description

g-select	g-range (g)	Sensitivity (mV/g)
0	1.5	800
1	6	206

$$C = A\varepsilon/D \tag{3.4}$$

where A is the area of the beam, ε is the dielectric constant, and D is the distance between the beams. Then, the acceleration information can be extracted from the sensor (Table 3.3).

There are two sensitivities that can be chosen by users:

3.13.2 Demonstration

1. **Components**

- DFRobot UNO R3 board and USB cable × 1.
- DFR0143 (MMA7361 triple axis accelerometer) × 1.
- Jumper wires × *n*.

2. **Hardware Setting**

The triple axis accelerometer has nine pins (three pins for each axis): *X*/*Y*/*Z*-Axis, VCC, and GND. *X*, *Y*, and *Z* should be connected to analog voltage outputs (here, A0, A1, and A2, respectively). The VCC should be connected to 5 V and the GND should be connected to a common ground (Fig. 3.40).

This sample shows how to measure the angle value using two axis values (*X* and *Z*):

3. **Sample Codes**

```
#include<math.h>
#include<stdio.h>
#define A_X 0 //A0,A1,A2 are used
#define A_Y 1
#define A_Z 2
double b;
void setup() {
  Serial.begin(9600);//configure baud rate to 9600 bps
}
void loop() {
  float a;
  int val_x = 0, val_y = 0, val_z = 0;
  for (int i = 0; i < 10; i++) {
    val_x += analogRead(A_X); delay(2);
    val_y += analogRead(A_Y); delay(2);
    val_z += analogRead(A_Z); delay(2);
  }
  val_x = val_x / 10; // noise removal
  val_y = val_y / 10;
  val_z = val_z / 10;
  delay(300);
  Serial.print("X_Axis: "); // output
  Serial.print(val_x);
  Serial.print("Z_Axis: ");
  Serial.print(val_z);
  Serial.print("    ");
  //320 is the analog output with horizontal state
  b = (double) (abs(val_x - 320)) / (abs(val_z - 320));
  Serial.print("B: ");
  Serial.print(b);
  Serial.print("    ");
  a = atan(b);
  Serial.print("A: ");
  Serial.println(a / 3.14 * 180); //the value of Angle
}
```

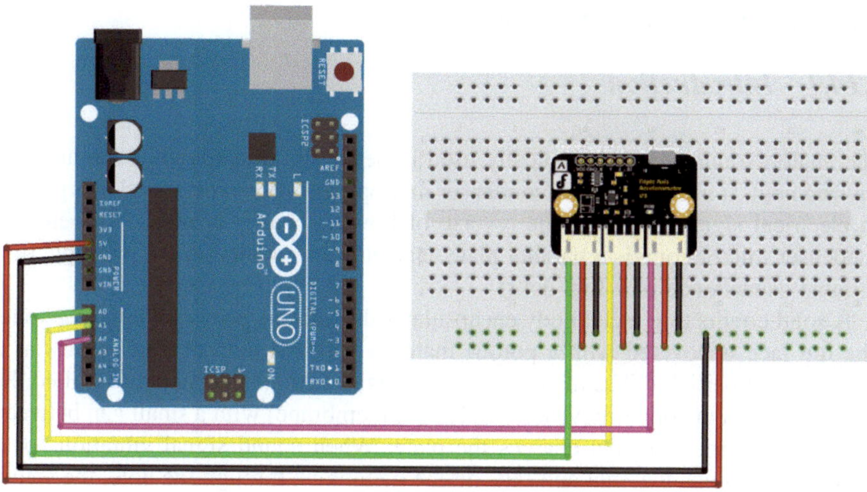

Fig. 3.40 A diagram of the layout of the DFR0143 and UNO R3

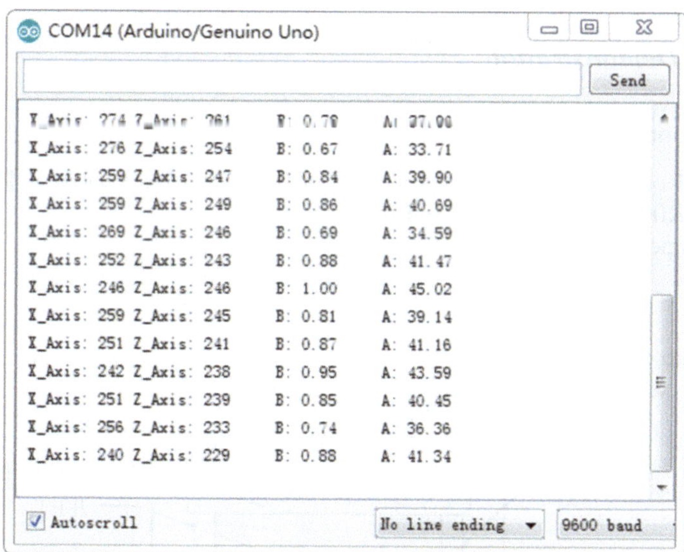

Fig. 3.41 Log messages from Arduino using DFR0143

4. **Results**

After uploading the program to your Arduino IDE, opening the serial monitor (Baudrate = 9600), and tilting the sensor, you should see values as follows (Fig. 3.41).

3.14 Analog Sound Sensor

3.14.1 Introduction

The analog sound sensor is typically used in detecting loud sounds in an ambient environment. It uses an electret condenser microphone to convert sound energy into an electrical signal, which is the reverse of how a speaker operates. The electret condenser microphone is a parallel plate capacitor and works on the principle of a variable capacitance (see Fig. 3.42).

A solid conducting metal body encapsulates the various parts of the microphone. The top face is covered with a porous material with the help of glue. It acts as a filter for dust particles. The microphone consists of two plates, one fixed (called the back plate) and the other moveable (called the membrane) with a small gap between them. An electric potential charges the plate. When sound signal vibrations pass through the porous material and falls on the membrane through the hole, the sound strikes the membrane and it starts moving, thereby changing the capacitance between the plates which in turn results in a variable electric current to flow.

3.14.2 Demonstration

1. **Components**
   - DFRobot Romeo BLE microcontroller board and USB cable × 1.
   - DFR0034 (analog sound sensor) × 1.
   - Jumper wires × 1.

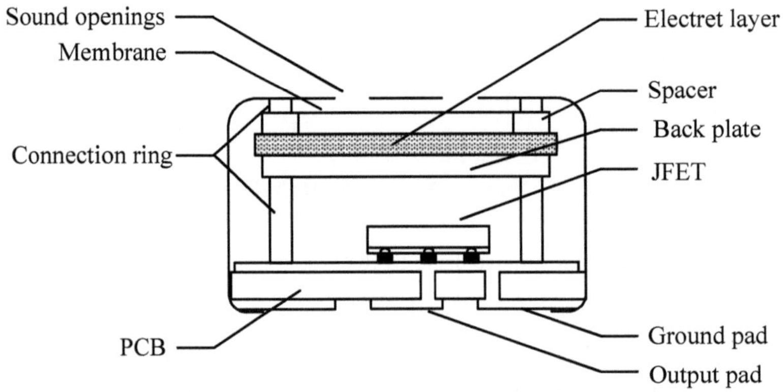

Fig. 3.42 The diagram of an electret condenser microphone

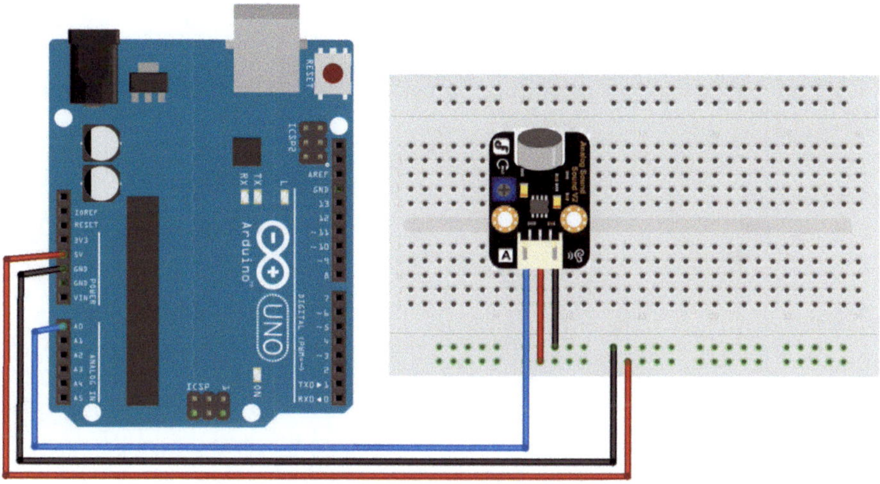

Fig. 3.43 A diagram of the layout of the DFR0034 and UNO R3

2. **Hardware Setting**

The analog sound sensor has three pins, Output, VCC, and GND. The Output should be connected to analog voltage outputs (here, PIN 0). The VCC should be connected to 5 V and the GND should be connected to a common ground on the DFRobot Romeo BLE board (Fig. 3.43).

3. **Sample Codes**

```
#define voicePin 0
void setup() {
  Serial.begin(9600); // configure baud rate to 9600 bps
}
void loop() {
  int val;
  val = analogRead(voicePin); //read  mic sensor
  Serial.println(val, DEC); //print the sound value
  delay(100);
}
```

4. **Results**

After uploading the program to your Arduino, opening the serial monitor (Baudrate = 9600), and generating some sounds, you should see values as follows (Fig. 3.44).

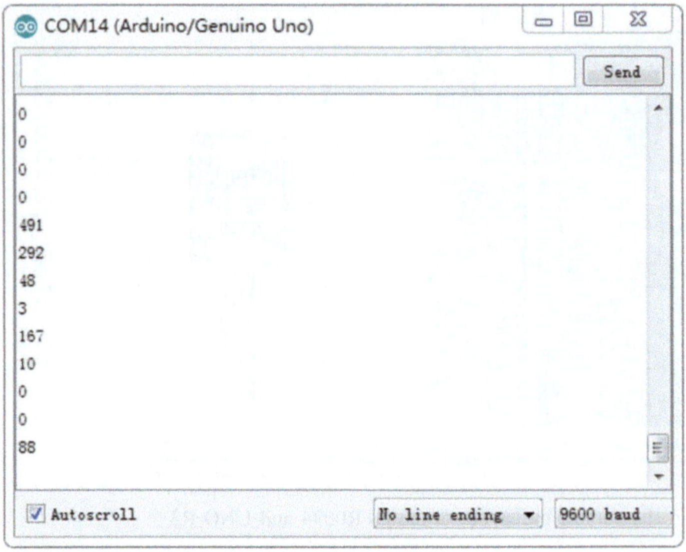

Fig. 3.44 Log messages from Arduino using DFR0034

3.15 Voice Recognition Module

3.15.1 Introduction

Voice recognition is a technology through which sounds, words, or phrases spoken by humans are converted into electrical signals. These signals are transformed into coding patterns that have been assigned meaning. It is widely used in home, health, public, and entertainment devices. The most common approaches to voice recognition can be divided into two classes: template matching and feature analysis. In this chapter, the voice recognition module we used belonged to the template matching class. The first step for the user is to speak a word or phrase into the microphone. The microphone transforms the sound signal to an electrical signal, which is then digitized by an analog-to-digital (A/D) converter, and is stored in the memory. To determine the meaning of this voice input, the module attempts to match the input with a digitized voice template, which is stored in the program. The module will match this template with the actual input using a simple conditional statement.

3.15.2 Demonstration

1. **Components**

- DFRobot UNO R3 board and USB cable × 1.
- DFR0177 (voice recognition module) × 1.
- LED × 1.
- Resistor (220 Ω) × 1.
- Jumper wires × *n*.

2. **Hardware Setting**

The voice recognition module is an expansion board of Arduino, which can be plugged into the UNO directly (Fig. 3.45).

DFR0177 is a Chinese module, so we can use Chinese pinyin to substitute English with similar pronunciation when writing codes. Table 3.4 shows some pinyin codes that may be useful in programming.

Fig. 3.45 A diagram of the layout of the DFR0177 and UNO R3

Table 3.4 Some English commands in Chinese pinyin

English	Chinese	English	Chinese
Start	si da te	End	en de
Up	a pu	Down	dang
Go	gou	Back	bai ke
Left	lai fu te	Right	ruai te
Lights on	lai ci ang	Lights off	lai ci ao fu

3. **Sample Codes**

Before we verify these codes with Arduino, the voice recognition library, i.e., "VoiceRecognition.h", should be downloaded first (https://github.com/tyjjr/voiceRecognition1.1/). Then, import the library in the Arduino IDE with the path as "Sketch → Include Library → Add.Zip Library."

```
#include <avr/wdt.h>
#include <VoiceRecognition.h>
VoiceRecognition Voice;
#define Led 13 // define the light digital pin
void setup() {
  Serial.begin(9600);
  pinMode(Led, OUTPUT);
  digitalWrite(Led, LOW);
  Voice.init();
  Voice.addCommand("lai ci on", 0);
  Voice.addCommand("lai ci ao fu", 1);
  Voice.start();
  wdt_enable(WDTO_1S);
}
void loop() {
  switch (Voice.read()) {
    case 0:
      digitalWrite(Led, HIGH);
      Serial.println("received 'lights on',command flag'0'");
      break;
    case 1:
      digitalWrite(Led, LOW);
      Serial.println("received 'lights off',command flag'1'");
      break;
  }
  wdt_reset();
}
```

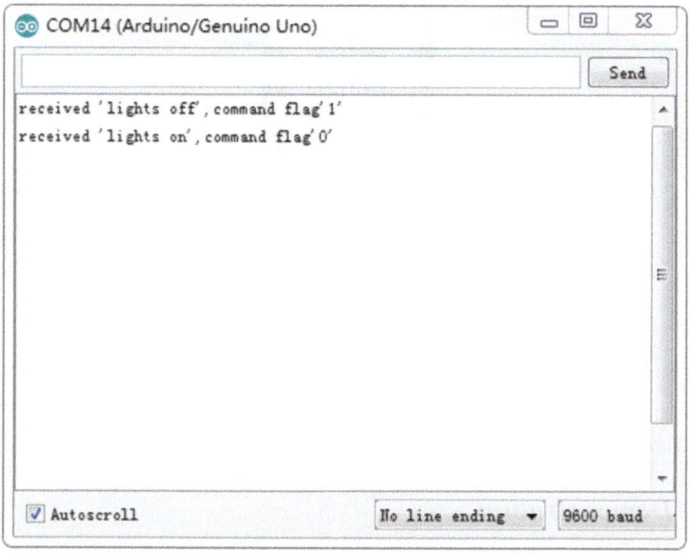

Fig. 3.46 Log messages from Arduino using DFR0177

4. **Results**

After uploading the program to your Arduino, the serial monitor (Baudrate = 9600) needs to be opened; the digital light illuminates after DFR0177 receives the voice "lights on." Further, the serial monitor shows the following (Fig. 3.46).

3.16 Digital Vibration Sensor

3.16.1 Introduction

Vibration detection is critical in security systems that are normally placed at window grills or doors. The DFRobot Digital Vibration Sensor is a digital plug that plays sensor blocks. It can sense weak vibration signals, which can be realized with the shock interaction with relevant works. The vibration sensor is made up of a vibration switch that contains a spring and a conductive material passing through (no contact) the center of the spring (Fig. 3.47).

The switch is in an open circuit OFF-state when static. When external force resulting in vibrations or movement speeds that produce adequate (partial) centrifugal force are applied, the spring touches the conductive material; then, the circuit is complete and the switch is in the ON-state. When the external force disappears, the switch goes back to the open circuit OFF-state.

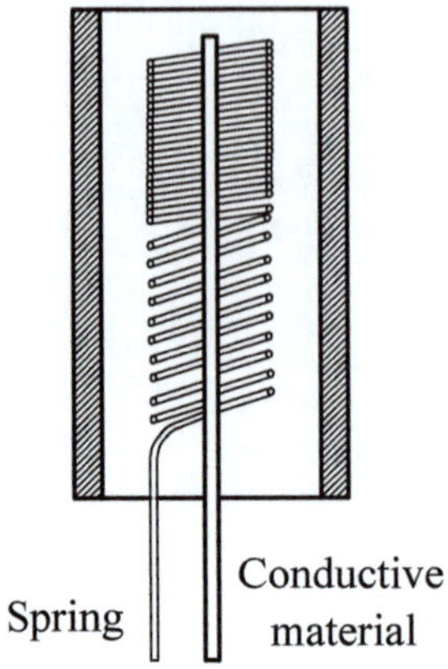

Fig. 3.47 The diagram of a vibration switch

3.16.2 Demonstration

1. **Components**

- DFRobot UNO R3 board and USB cable × 1.
- DFR0027 (digital vibration sensor) × 1.
- LED × 1.
- Resistor (220 Ω) × 1.
- Jumper wires × *n*.

2. **Hardware Setting**

The vibration sensor has three pins: VCC, GND, and Output. The VCC should be connected to 5 V and the GND should be plugged into the common ground. The Output pin should be connected to the digital pin (here, PIN 3) (Fig. 3.48).

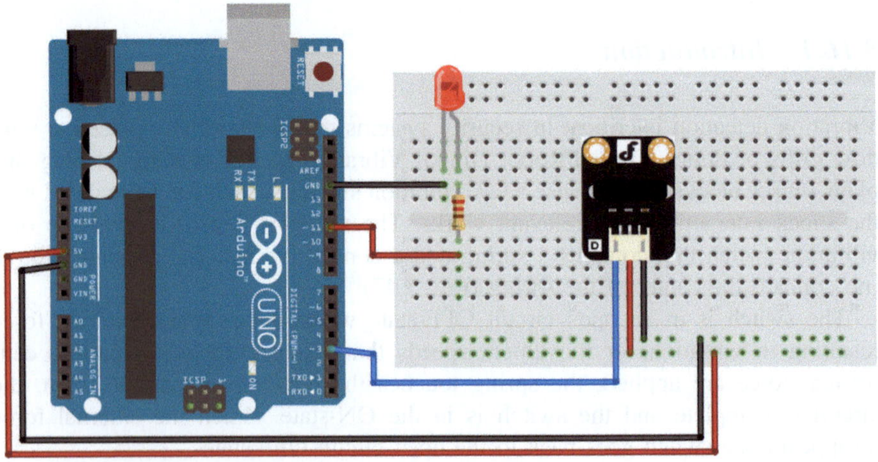

Fig. 3.48 A diagram of the layout of the DFR0027 and UNO R3

3. **Sample Codes**

```
#define ledPin 13 // define the light digital pin
#define vibrationPin 3 // define the vibration digital pin
unsigned char state = 0;
void setup() {
  pinMode(ledPin, OUTPUT);
  pinMode(vibrationPin, INPUT);
  attachInterrupt(1, blink, FALLING); // select interrupt 1, the
service routine is blink
}
void loop() {
  if (state != 0) {
    state = 0;
    digitalWrite(ledPin, HIGH); //Trigger the blink function when
the falling edge is detected
    delay(500);
  }
  else
    digitalWrite(ledPin, LOW);
}
// interrupt service routine
void blink()  {
  state++;  // change the state
}
```

4. **Results**

After uploading the program to your Arduino, the digital light illuminates when DFR0027 vibrates. Of course, the light is off when the vibration disappears.

3.17 Flame Sensor

3.17.1 Introduction

The flame sensor can be used to detect fire or other wavelengths at 760–1100 nm light. Small flames like a lighter flame can be detected at roughly 0.8 m. The detection angle is roughly 60° and the sensor is particularly sensitive to the flame spectrum. In the fire-fighting robot game, the flame sensor plays an important role

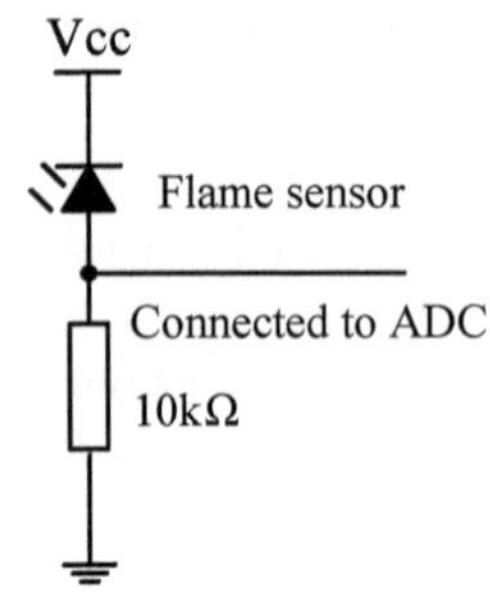

Fig. 3.49 The wiring diagram of flame sensor

in the probe, which can be used as a robot's eyes to find a fire source or football. The sensor can be employed in fire-fighting robots and soccer robots. The circuit of the flame sensor is as follows (Fig. 3.49):

The negative lead is connected to 5 V and the positive lead is connected to the GND through a 10 KΩ resistor. When the intensity of infrared light changes, the resistance of the sensor changes too, which then leads to a change of voltage.

The flame sensor's operating temperature is −25 to 85 °C. Further, it should be noted that the probe distance from the flame should not be too near in order to avoid damage. The shortest test distance is 80 cm, if the flame is bigger, it should be tested at a farther distance.

3.17.2 Demonstration

1. **Components**

   - DFRobot UNO R3 board and USB cable × 1.
   - DFR0076(Flame sensor) × 1.
   - LED × 1.
   - Resistor (220 Ω) × 1.
   - Jumper wires × *n*.

2. **Hardware Setting**

The flame sensor has three pins: VCC, GND, and Output. The VCC should be connected to 5 V and the GND should be plugged into the common ground. The Output pin should be connected to the analog pin (here, A0) (Fig. 3.50).

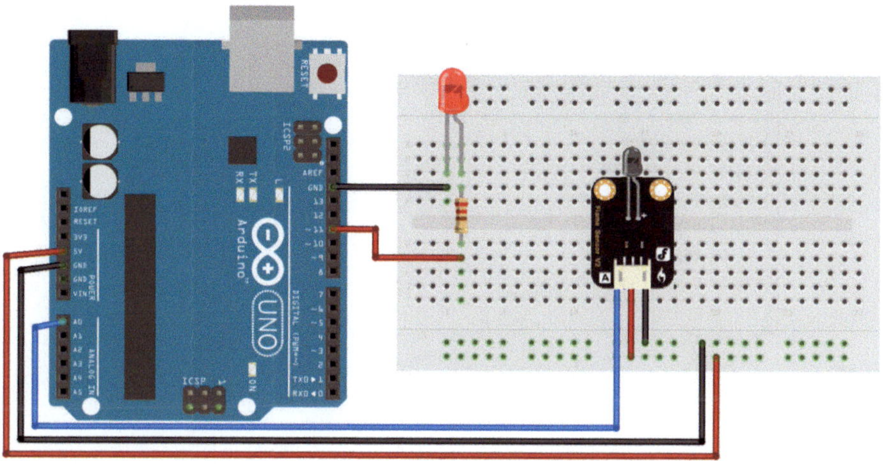

Fig. 3.50 A diagram of the layout of the DFR0076 and UNO R3

3. **Sample Codes**

```
#define flamePin A0
void setup() {
  Serial.begin(9600); // configure baud rate to 9600 bps
}
void loop() {
  Serial.print("Sensor Value: ");
  int val = analogRead(flamePin);
  Serial.println(val);
  if (val < 800) {
    Serial.println(" FLAME DETECTED!");
  }
  else {
    Serial.println();
  }
  delay(100);
}
```

4. **Results**

When the flame sensor detects a flame, the data is observed on the serial monitor (Fig. 3.51).

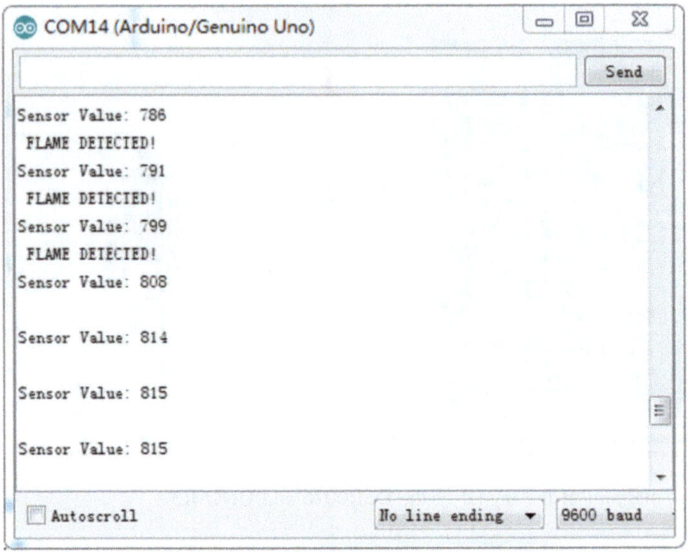

Fig. 3.51 Log messages from Arduino using DFR0076

3.18 Capacitive Touch Sensor

3.18.1 Introduction

In a number of cases, machines, home appliances, and electronic devices have to be controlled by human beings. It is part of everyone's daily life and we are familiar with switches, push buttons, keyboards, knobs, and slider controls. Over the last few years, a new way to control device has emerged. We are referring to the touch sensor. The reasons for this development are durability, robustness, attractive product design, and cost. Touch sensors, unlike mechanical devices, do not contain moving parts. Hence, they are more durable than mechanical input devices. Touch sensors are robust as there are no openings for humidity and dust to enter.

When any object with capacitive characteristics—such as a finger—comes close to a capacitive touch sensor, it acts as another capacitor due to its dielectric nature. This varies the effective capacitance of the system, which is used to detect the touch.

The finger acts as one of the parallel plates, while another parallel plate is connected to the sensor input of the chip, as shown in the illustration below (Fig. 3.52).

p.s. (1) C_F: finger capacitance, C_S: sensor capacitance, *d*: distance between the plates
(2) Untouched sensor pad with parasitic capacitance C_S, touched sensor pad with additional touch capacitance C_F

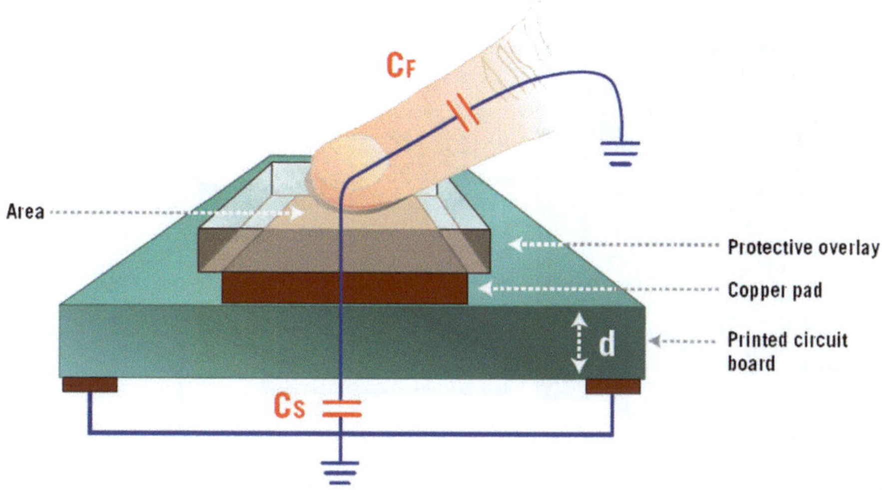

Fig. 3.52 The principle of capacitive touch sensor

The iron content in human blood creates strings of capacitors that are aligned to the surface of the body. When such strings of capacitors come in proximity with a conductor, a capacitance that is essentially coupled to the ground is created, which causes a change in the measured voltage, determining the touch.

3.18.2 Demonstration

1. **Components**

- DFRobot UNO R3 board and USB cable × 1.
- DFR0030 (Touch sensor) × 1.
- LED × 1.
- Resistor (220 Ω) × 1.
- Jumper wires × 1.

2. **Hardware Setting**

The touch sensor has three pins: VCC, GND, and Output. The VCC should be connected to 5 V and the GND should be plugged into the common ground. The Output pin should be connected to the digital pin (here, PIN 2) (Fig. 3.53).

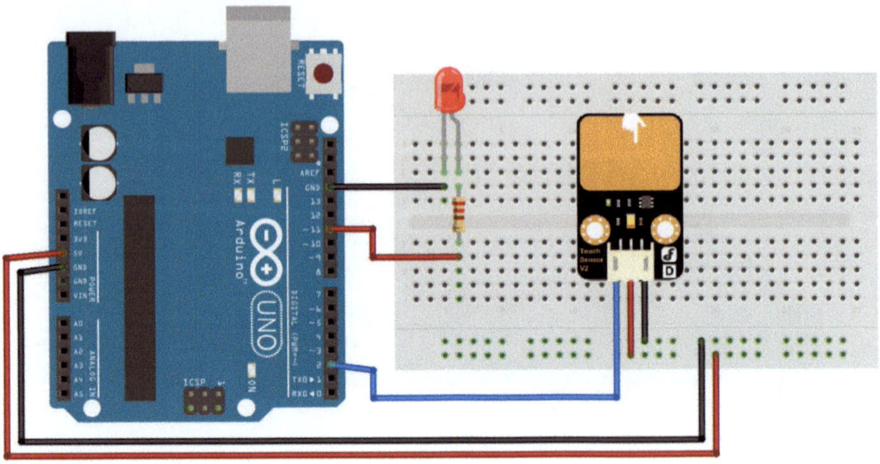

Fig. 3.53 A diagram of the layout of the DFR0030 and UNO R3

3. **Sample Codes**

```
int ledPin = 13; // Connect LED on digital pin 13
int keyPin = 2; // Connect Touch sensor on digital Pin 2
void setup() {
  pinMode(ledPin, OUTPUT); // Set ledPin to output mode
  pinMode(keyPin, INPUT); // Set touch sensor pin to input mode
}
void loop() {
  // Read Touch sensor signal
  if (digitalRead(keyPin) == HIGH) {
    // if Touch sensor is HIGH, then turn on
    digitalWrite(ledPin, HIGH);
  }
  else {
    // if Touch sensor is LOW, then turn off
    digitalWrite(ledPin, LOW);
  }
}
```

4. **Results**

After uploading the program to your Arduino, the digital light illuminates when your finger or metal object touches the metal surface of the transducer.

Chapter 4
Electromechanical Control Using the Arduino

4.1 DC Motor

4.1.1 Overview

DC motors are becoming increasingly common in a variety of motor applications such as fans, pumps, appliances, automation, and automotive drive. The reasons for their increased popularity are better speed versus torque characteristics, high efficiency, long operating life, and noiseless operation. In addition to these advantages, the ratio of torque delivered to the size of the motor is higher, making it useful in applications where space and weight are critical factors. This makes them attractive options for designers who are interested in robotics. You can run a DC motor by supplying a voltage difference across its leads. However, you need to overcome certain challenges in order to drive them effectively. The most common goals are variable speed and direction. To control the direction of the spin of DC motors, without changing the way that the leads are connected, an H-Bridge is commonly used.

An H-bridge is an electronic circuit that can drive the motor in both directions. As shown in Fig. 4.1, one H-bridge uses four transistors connected in such a way that the schematic diagram appears like an "H".

The basic operating mode of an H-bridge is fairly simple: if Q1 and Q3 are turned on, the left lead of the motor is connected to the power supply, while the right lead is connected to the ground. Current starts flowing through the motor, which energizes the motor in (let us say) the forward direction and the motor shaft starts spinning.

If Q4 and Q2 are turned on, the reverse happens, the motor gets energized in the reverse direction, and the shaft will start spinning backwards.

T. Pan and Y. Zhu, *Designing Embedded Systems with Arduino*,
DOI 10.1007/978-981-10-4418-2_4

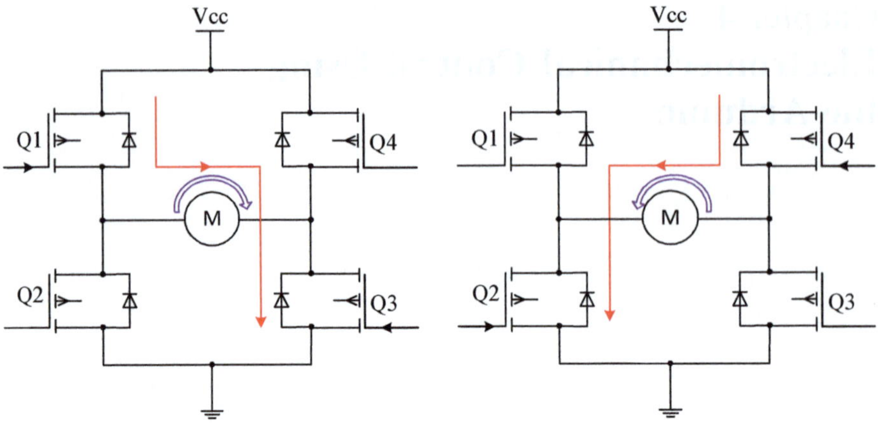

Fig. 4.1 H-bridge and the direction of motor

In a bridge, you should never ever close both Q1 and Q2 (or Q3 and Q4) at the same time. If you did that, you may create a really low-resistance path between the power and GND, effectively short-circuiting your power supply. This condition is called "shoot-through" and is an almost guaranteed way to quickly destroy your bridge, or something else in your circuit.

4.1.2 Driven Circuit Design

In a number of cases, especially for little toy motors, you do not need to build a whole H-bridge circuit from scratch. In fact, using a chip can save you a considerable amount of trouble with regard to offset voltages; if you have a different motor supply voltage than your logic voltage, you will need drivers between the logic and the power transistors. There are several packaged IC chips (such as L293D, L298N, TA7257P, SN754410, etc.) that are inexpensive and easy to build into a circuit.

The common L293N dual motor controller is a 16-DIP IC chip that contains two protected driver circuits capable of delivering up to 600 mA of continuous current to each motor at up to 36 VDC (see Fig. 4.2). L298N is a similar chip that can deliver 2 amps to each motor. These chips (and others) accept standard 0–5 V input signals and have internal logic gates to prevent accidental overloading and commanding the controller into a destructive state.

Notice in Fig. 4.2 that there are six pins labeled IN1, IN2, IN3, IN4, ENA, and ENB. You can use digital pins on the Arduino to control the four input pins and set the motor direction, while using a PWM signal on each enable pin (ENA and ENB) to set the speed of each motor. It should be pointed out that L298N does not have

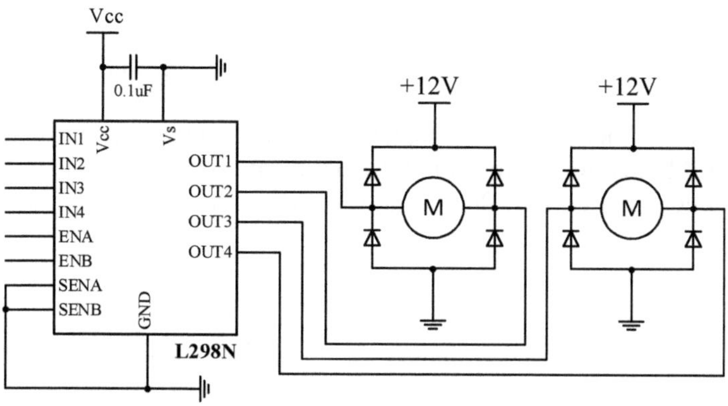

Fig. 4.2 Schematic of L298N driven circuit

Table 4.1 Motor direction control

IN1	IN2	Action
LOW	LOW	Motor breaks and stops
HIGH	LOW	Motor turns forward
LOW	HIGH	Motor turns backward
HIGH	HIGH	Motor breaks and stops

built-in protection diodes, so you will need to add those. The datasheet for the L298N specifies "fast recovery" 1-amp diodes; an inexpensive selection is the 1N4933, available from most online electronic parts outlets.

Let us look at how to control just one of the motors, Motor1. In order to activate the motor, the pin ENA must be high. You then control the motor and its direction by applying a low or high signal to the Input1 and Input2 lines, as shown in Table 4.1.

The L298N H-bridge module can be used with motors that have a voltage of between 5 and 35 V DC.

4.1.3 Demonstration

1. **Components**

- DFRobot Romeo board and USB cable × 1
- DC motor × 1
- 9 V battery × 1
- Jumper wires × *n*

2. **Hardware Setting**

Connect four motor wires to motor terminal. Then, apply power through the motor power terminal (Fig. 4.3).

The PWM DC motor control is implemented by manipulating two digital IO pins and two PWM pins. As illustrated in the diagram above (Fig. 4.3), Pins 4 and 7 are motor direction control pins, Pins 5 and 6 are motor speed control pins (shown in Table 4.2).

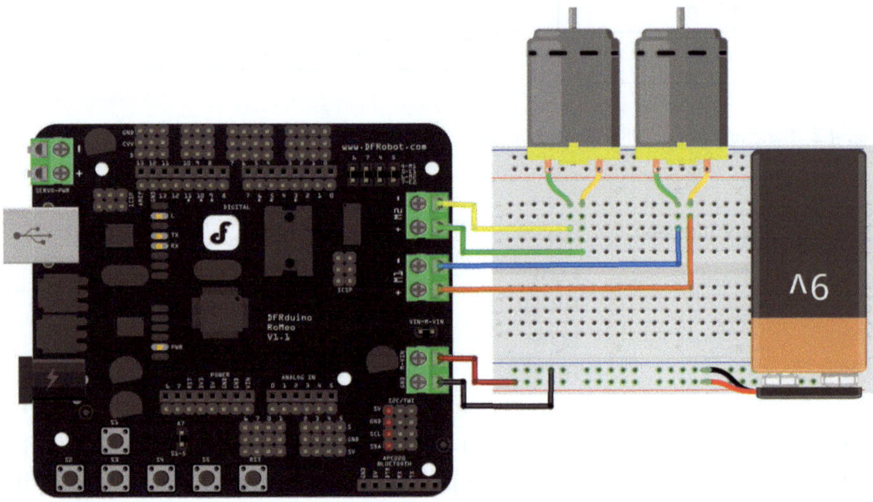

Fig. 4.3 Romeo motor connection diagram

Table 4.2 PWM mode

Pin	Function
Digital 4	Motor 1 direction control
Digital 5	Motor 1 PWM control
Digital 6	Motor 2 PWM control
Digital 7	Motor 2 direction control

3. Sample Codes

```
#define keyPin A0;  // analogy key input
int E1 = 5;  //M1 Speed Control
int E2 = 6;  //M2 Speed Control
int M1 = 4;  //M1 Direction Control
int M2 = 7;  //M2 Direction Control
unsigned long sampleTime;
unsigned int SampleInterval = 100;    // sampling time is 0.1s
int adc_key_val[5] = {50, 200, 400, 600, 800 }; // threshold for
comparison
boolean blnKey; // flag for against key shake
int NUM_KEYS = 5; // 5 keys
int adc_key_in;
int key = -1;

void setup() {
  Serial.begin(9600); //Configure baud rate 9600
}

void loop() {
  if (millis() - sampleTime >= SampleInterval) {
    sampleTime = millis();
    adc_key_in = analogRead(A0);    // // read the key value
    if (adc_key_in < 1000) { // if key press
      if (blnKey == 0)
        blnKey = 1; // wait for debounce time
      else
        key = get_key(adc_key_in);   // convert into key press
    }
    else {
      key = -1;   // no key press
      blnKey = 0;
    }
  }
  if (key >= 0) {
    switch (key) {
      case 0: {
          Serial.println("S1 OK");
          advance (100, 100);  //move forward
          break;
        }
      case 1: {
```

```
            Serial.println("S2 OK");
            back_off (100, 100); // move backward
            break;
          }
        case 2: {
            Serial.println("S3 OK");
            turn_L (100, 100); // turn left
            break;
          }
        case 3: {
            Serial.println("S4 OK");
            turn_R (100, 100); // turn right
            break;
          }
        case 4:  {
            Serial.println("S5 OK");
            stop(); // stop
            break;
          }
      }
    }
}

// convert ADC value to key number
int get_key(unsigned int input) {
  int k;
  for (k = 0; k < NUM_KEYS; k++) {
    if (input < adc_key_val[k]) {   // compared the threshold
      return k;
    }
  }
  if (k >= NUM_KEYS)
    k = -1;  // no valid key pressed
  return k;
}

void stop(void) {  //Stop
  digitalWrite(E1, LOW);
  digitalWrite(E2, LOW);
}
void advance(char a, char b) {  //Move forward
```

```
  analogWrite (E1, a);      //PWM Speed Control
  digitalWrite(M1, HIGH);
  analogWrite (E2, b);
  digitalWrite(M2, HIGH);
}
void back_off (char a, char b) {  //Move backward
  analogWrite (E1, a);
  digitalWrite(M1, LOW);
  analogWrite (E2, b);
  digitalWrite(M2, LOW);
}
void turn_L (char a, char b) {  //Turn Left
  analogWrite (E1, a);
  digitalWrite(M1, LOW);
  analogWrite (E2, b);
  digitalWrite(M2, HIGH);
}
void turn_R (char a, char b) {  //Turn Right
  analogWrite (E1, a);
  digitalWrite(M1, HIGH);
  analogWrite (E2, b);
  digitalWrite(M2, LOW);
}
```

4. **Analysis**

In this example, the PWM DC motor control is implemented by manipulating two digital IO pins and two PWM pins. Furthermore, the move direction is controlled by the button-on-board: "S0—move forward", "S1—move backward", "S2—turn left", "S3—turn right", "S4—stop" (Fig. 4.4).

The schematic of button-on-board is shown as follows.

It can be seen that if no button is pressed, the voltage of Pin A0 is 5 V (1023). If any button is pressed, the voltage of Pin A0 is less than 5 V and the specific values are shown in Table 4.3.

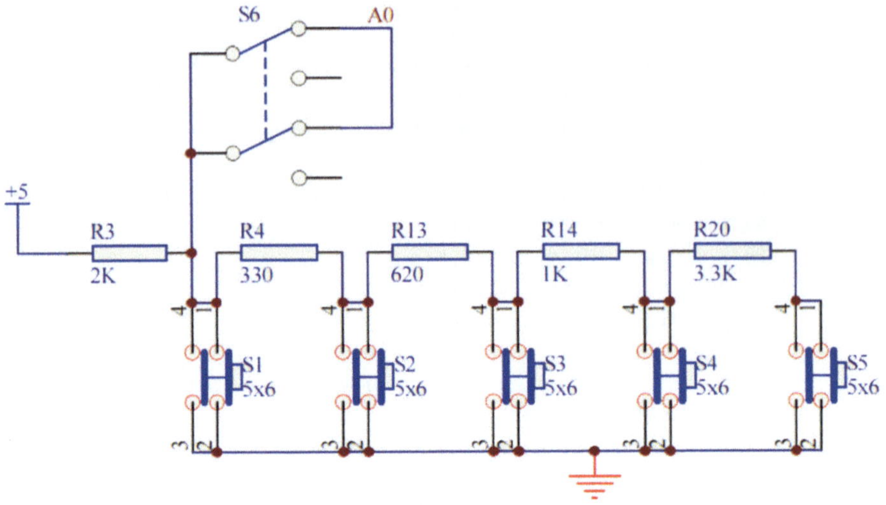

Fig. 4.4 Schematic of button-on-board of Romeo board

Table 4.3 Voltage of A0 when button is pressed

Button pressed	Voltage (V)	ADC value
S1	0	0
S2	0.7	145
S3	1.6	330
S4	2.5	505
S5	3.6	741

4.2 Stepper Motor

4.2.1 Overview

A stepper motor is a brushless, synchronous electric motor that converts digital pulses into mechanical shaft rotations. Each rotation of a stepper motor is divided into a set number of steps. The stepper motor must send a separate pulse for each step. The stepper motor can only receive one pulse and take one step at a time and each step must be of the same length. Since each pulse results in the motor rotating at a precise angle, you can precisely control the position of the stepper motor without any feedback mechanism (an open-loop controller).

As the digital pulses from the controller increase in frequency, the stepping movement converts into a continuous rotation with the velocity of the rotation directly proportional to the frequency of the control pulses. Stepper motors are widely used because of their low cost, high reliability, and high torque at low speeds. Furthermore, an energized stepper motor maintains full torque at a standstill position. Their rugged construction enables you to use stepper motors in a wide environmental range.

4.2.2 Working Principle of Stepper Motor

A stepper motor is constructed from ferromagnetic material with salient poles as shown in Fig. 4.5. The stator is made from a stack of steel laminations and has six equally spaced projection poles (or teeth), each wound with an excitation coil. The rotor, which maybe solid or laminated, has four teeth of the same width. As seen, there are three independent stator circuits or phases A, B, and C and each one can be energized by a direct current pulse from the drive circuit.

A simple circuit arrangement for supplying current to the stator coils in proper sequence is shown in Fig. 4.6. The six stator coils are connected in two-coil groups to form three separate circuits called phases. Each phase has its own independent switch. Diametrically, opposite pairs of stator coils are connected in series such that when one tooth becomes an *N*-pole, the other becomes an *S*-pole. Although shown as mechanical switches in Fig. 4.5, in practice, switching of phase currents is performed with the help of solid-state control. When there is no current in the stator coils, the rotor is completely free to rotate. Energizing one or more stator coils causes the rotor to step forward (or backward) to a position that forms a path of least resistance with the magnetized stator teeth. The step angle of this three-phase, four-rotor teeth motor is $\beta = \frac{360}{4\times 3} = 30^{\circ}$

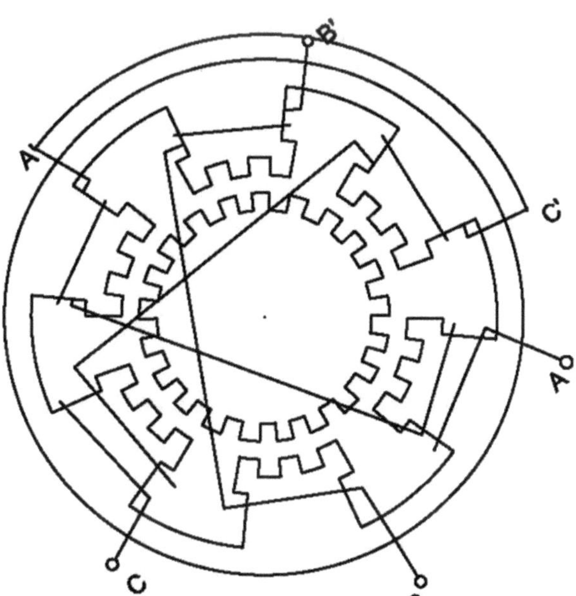

Fig. 4.5 Structure of stepper motor

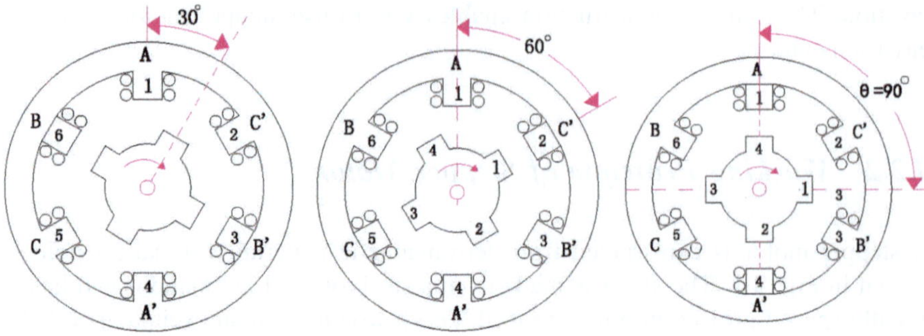

Fig. 4.6 Sequence of supplying current to stator coil

Table 4.4 One-phase-ON mode "A-B-C-A"

A	B	C	
+	0	0	0°
0	+	0	30°
0	0	+	60°
+	0	0	90°

In general, there are three kinds of phase current switching modes, i.e., one-phase-ON, two-phase-ON, and half-step operation.

As shown in Fig. 4.6, energizing stator phases in sequence A-B-C-A, the rotor will rotate clockwise in 30° steps. If the switch sequence is made C-B-A-C, the rotor will rotate anticlockwise. This mode of operation is known as the one-phase-ON mode operation and is the simplest and most widely used way of making the motor step. The stator phase switching truth table is shown in Table 4.4.

In Fig. 4.7, If the stator phases are switched in the sequence AB-BC-CA-AB (shown in Table 4.5), the motor will take full steps of 30° each (as in the one-phase-ON mode), but its equilibrium positions are interleaved between the full-step positions.

If the stator phases are excited in the sequence A-AB-B-BC-CA-A (shown in Table 4.6), i.e., alternately in the one-phase-ON and two-phase-ON modes, then it is sometimes known as "wave" excitation and it causes the rotor to advance in steps of 15°.

4.2.3 *Driven Principle of Stepper Motor*

The stepper motors can be split into two types: unipolars and bipolars. The type of the motor is important to determine a compatible electronic device. In general, a stepper motor is controlled by a series of electromagnetic coils surrounding the shaft and designed to convert the electrical pulse in mechanical movements.

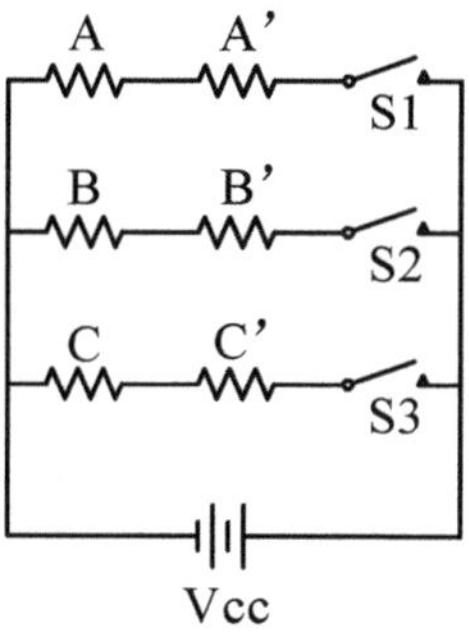

Fig. 4.7 Principle of phase current switch

Table 4.5 Two-phase-ON mode "AB-BC-CA-AB"

A	B	C	
+	+	0	15°
0	+	+	45°
+	0	+	75°
+	+	0	105°

Table 4.6 Half-step operation one-phase-ON and two-phase-ON mode "A-AB-B-BC-C-CA-A"

	A	B	C	
A	+	0	0	0°
AB	+	+	0	15°
B	0	+	0	30°
BC	0	+	+	45°
C	0	0	+	60°
CA	+	0	+	75°
A	+	0	0	90°

A unipolar stepper motor has one winding with a center tap per phase. Each section of windings is switched on for each direction of magnetic field. Since in this arrangement, a magnetic pole can be reversed without switching the direction of the current, the commutation circuit can be made very simple (e.g., a single transistor) for each winding. Typically, given a phase, the center tap of each winding is made common: giving three leads per phase and six leads for a typical two-phase motor. Often, these two phase commons are internally joined, so the motor has only five leads. The circuit for a unipolar stepper motor is shown in Fig. 4.8.

A microcontroller or stepper motor controller can be used to activate the drive transistors in the right order, and this ease of operation makes unipolar motors popular with hobbyists; they are probably the cheapest way to obtain precise angular movements.

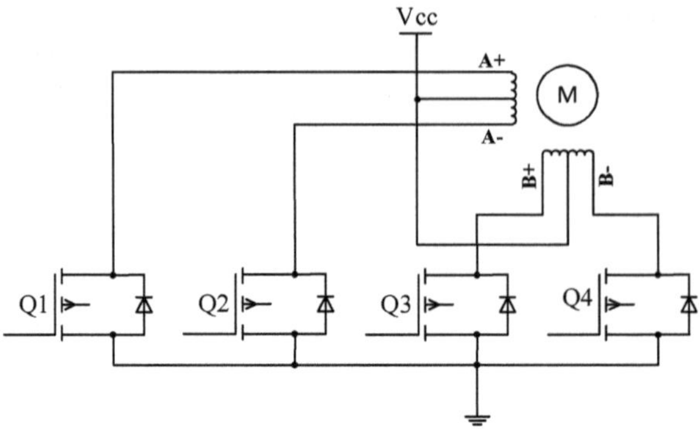

Fig. 4.8 Driven circuits for unipolar stepper motor

Bipolar motors have a single winding per phase. The current in a winding needs to be reversed in order to reverse a magnetic pole, so the driving circuit must be more complicated, typically with an H-bridge arrangement (shown in Fig. 4.9). There are two leads per phase; none are common. Because windings are better utilized, they are more powerful than a unipolar motor of the same weight. This is owing to the physical space occupied by the windings.

Though a bipolar stepper motor is more complicated to drive, the abundance of driver chips means that this is much less difficult to achieve.

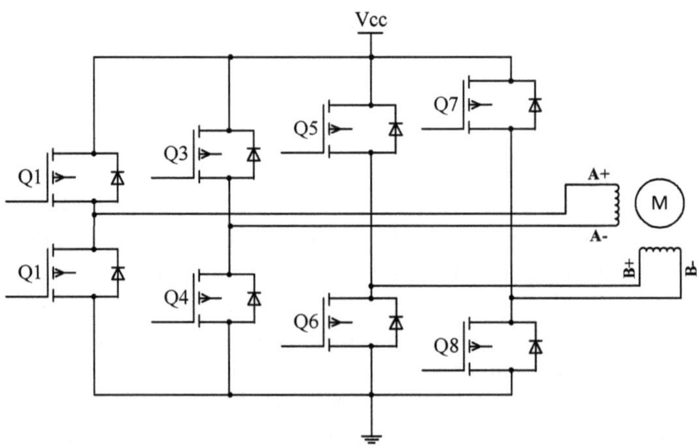

Fig. 4.9 Driven circuits for bipolar stepper motor

4.2.4 Driven Circuit Design

There are actually many ways you can interface a stepper motor to your controller; out of them the most used interfaces are:

- Interface using L293D—H-Bridge Motor Driver
- Interface using ULN2003/2004—Darlington Arrays

1. **Connoting bipolar stepper motor using L293D**

L293D contains two H-bridges for driving stepper motors. One L293D can, in theory, drive one bipolar two-phase stepper motor, if you supply the correct sequence.

L293D IC has 16 pins. Here is how each of the pins should be connected:

Pins 1, 9 Enable pins. Hook them together and you can either keep them high and run the motor all the time or control them with you own controller.

Pins 3, 6, 11, 14 Here is where you plug in the two coils. To determine which wires correspond to each coil, you can use a multimeter to measure the resistance between the wires. The wires that correspond to the same coil have a much lower resistance than wires that correspond to different coils. Then, you can get one coil hooked up to Pins 3 and 6 and another one hooked up to Pins 11 and 14.

Pins 4, 5, 12, 13 These are attached to ground.

Pin 8 The motor voltage, for the motors we are using, is 12 V.

Pin 16 +5 V. It is the power supply of the chip and it is a good idea to keep this power supply separate from your motor power.

Pins 2, 7, 10, 15 Control signals. Here is where you supply the pulse sequence. The following is how you pulse them for a single cycle (to move the motor in the opposite direction, just reverse the steps. i.e., from step 4 to step 1).

The driving circuit is shown in Fig. 4.10.

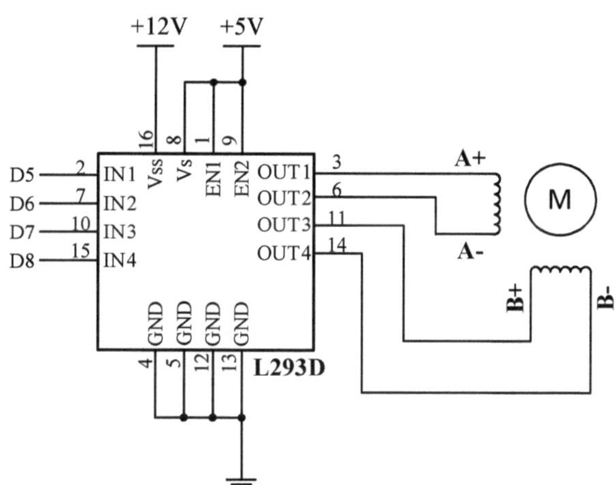

Fig. 4.10 Driving one stepper motor using L293D

As you see in the circuit, above the four Pins "D5, D6, D7, and D8" control the motion and direction of the stepper motor according to the step sequence programmed in the controller.

2. **Connecting unipolar stepper motor using ULN2003/2004**

ULN2003/2004 internally employs high-voltage, high-current Darlington arrays each containing seven open collector Darlington pairs with common emitters. Each channel is rated at 500 mA and can withstand peak currents of 600 mA. Suppression diodes are included for inductive load driving and the inputs are pinned opposite the outputs to simplify board layout. ULN2003A is of 5 V TTL, CMOS. These versatile devices are useful for driving a wide range of loads including solenoids, relays, DC motors, LED displays, filament lamps, thermal printheads, and high-power buffers. ULN2003A are supplied in 16 pin plastic DIP packages with a copper leadframe to reduce thermal resistance.

The driven circuit is shown in Fig. 4.11. The Pins 1 to 4 of ULN2003 are connected to the microcontroller pins. Through Pins 13 to 16 of ULN 2003, stepper motor draws the control sequence for the sake of rotation. Pin 9 is common and Pin 8 is ground. The sequence 0011 (03H), 0110 (06H), 1100 (0CH), 1001 (09H) provides maximum torque in a two-phase configuration. This sequence energizes two adjacent phases, which offers an improved torque-speed product and greater holding torque.

4.2.5 Demonstration 1

1. **Components**

- DFRobot UNO R3 board and USB cable × 1
- 28BYJ48 Stepper motor × 1
- ULN2003A × 1
- Jumper wires × *n*

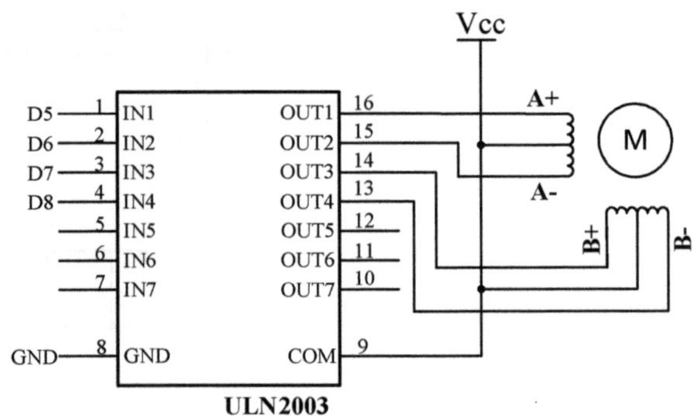

Fig. 4.11 Driving one stepper motor using ULN2003

2. **Hardware Setting**

In this example, the driven chip ULN2003 is used, the connecting diagram is shown in Fig. 4.12. This is the simplest way of interfacing a unipolar stepper to Arduino.

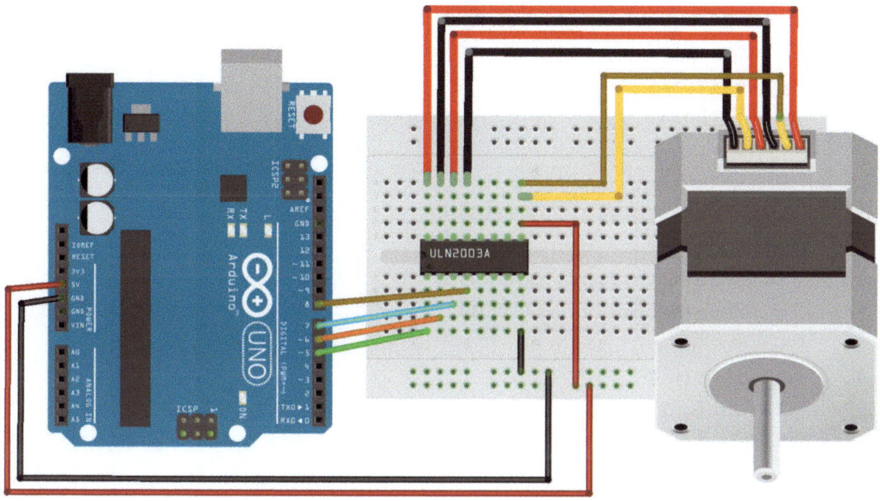

Fig. 4.12 UNO board and driven circuit diagram

2. Sample Codes

```
// Controlling a servo position using a potentiometer (variable
resistor)
int motorPin1 = 2; // input 1 of the stepper
int motorPin2 = 3; // input 2 of the stepper
int motorPin3 = 4; // input 3 of the stepper
int motorPin4 = 5; // input 4 of the stepper
int delayTime = 500;
int stepDelay = 25; // Delay between steps in milliseconds

void setup() {
  pinMode(motorPin1, OUTPUT);
  pinMode(motorPin2, OUTPUT);
  pinMode(motorPin3, OUTPUT);
  pinMode(motorPin4, OUTPUT);
}
void loop() {
  direction1();
}

void direction1() {
  A();
  BB();
  AA();
  B();
}

void direction2() {
  B();
  AA();
  BB();
  A();
}

void A() {
  digitalWrite(motorPin1, HIGH);
  digitalWrite(motorPin2, LOW);
  digitalWrite(motorPin3, LOW);
  digitalWrite(motorPin4, LOW);
  delay(delayTime);
}
```

```
void AA() {
  digitalWrite(motorPin1, LOW);
  digitalWrite(motorPin2, HIGH);
  digitalWrite(motorPin3, LOW);
  digitalWrite(motorPin4, LOW);
  delay(delayTime);
}
void BB() {
  digitalWrite(motorPin1, LOW);
  digitalWrite(motorPin2, LOW);
  digitalWrite(motorPin3, LOW);
  digitalWrite(motorPin4, HIGH);
  delay(delayTime);
}
void B() {
  digitalWrite(motorPin1, LOW);
  digitalWrite(motorPin2, LOW);
  digitalWrite(motorPin3, HIGH);
  digitalWrite(motorPin4, LOW);
  delay(delayTime);
}
```

4.2.6 Demonstration 2

1. **Components**

- DFRobot UNO R3 board and USB cable × 1
- 42BYGH1861A-C Stepper motor × 1
- DRI0023 (Stepper Motor Shield) × 1
- Jumper wires × *n*

2. **Hardware Setting**

In this example, the stepper motor shield (DRI0023) is used, which can be plugged into the UNO directly. Using this stepper motor shield, you can easily drive two stepper motors via just four digital I/Os (digital pins: 4, 5, 6, 7) (Fig. 4.13).

3. **Sample Codes**

```
int M1dirpin = 4; // define the motor digital pin
int M1steppin = 5;
int M2dirpin = 7;
int M2steppin = 6;
void setup() {
  pinMode(M1dirpin, OUTPUT);
  pinMode(M1steppin, OUTPUT);
  pinMode(M2dirpin, OUTPUT);
  pinMode(M2steppin, OUTPUT);
}
void loop() {
  delayMicroseconds(2);
  digitalWrite(M1dirpin, LOW);
  digitalWrite(M2dirpin, LOW);
  for (int j = 0; j <= 5000; j++) {
    digitalWrite(M1steppin, LOW);
    digitalWrite(M2steppin, LOW);
    delayMicroseconds(2);
    digitalWrite(M1steppin, HIGH);
    digitalWrite(M2steppin, HIGH);
    delay(1);
  }
}
```

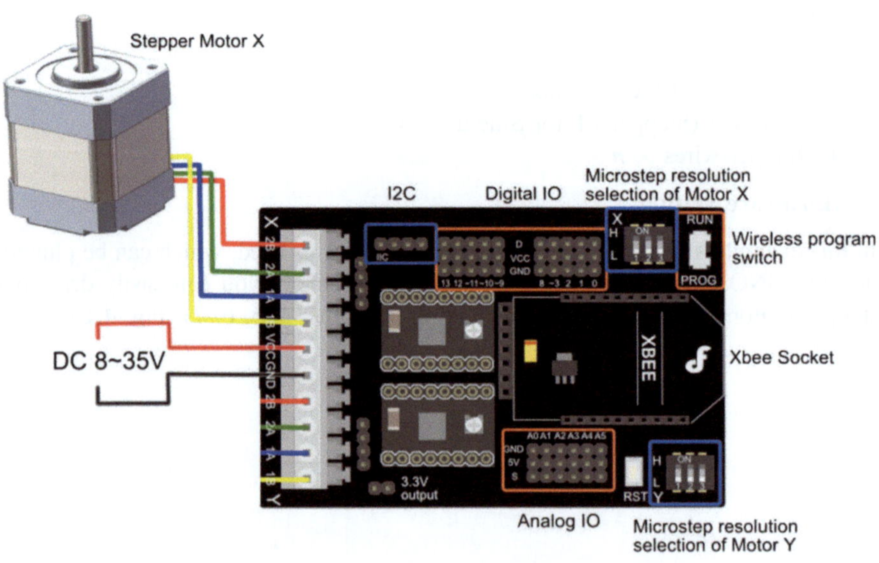

Fig. 4.13 Stepper motor and Romeo board connection diagram

4.3 Servo Motor

4.3.1 Overview

Servo motors are small controllable motors that tend to be implemented in many applications. There are servos with many different speeds, sizes, and torque capabilities, but all have three wires, power, ground, and control. Servo motors are popular with hobbyists because they are inexpensive, $15–$100, and control of servo motors with microcontrollers is universal for all models. Servos receive pulse width modulated (PWM) signals to determine in which manner to move.

A servo motor consists of several main parts, the motor and gearbox, a position sensor, an error amplifier and motor driver, and a circuit to decode the requested position. Figure 4.14 contains a block diagram of a typical servo motor unit.

At the initial position of a servo motor shaft, the position of the potentiometer knob is such that there is no electrical signal generated at the output port of the potentiometer. This output port of the potentiometer is connected to one of the input terminals of the error detector amplifier. Another input terminal of the error detector amplifier is connected to an external source. Then, the difference between these two signals is amplified in the error detector amplifier to feed the DC motor. This amplified error signal acts as the input power of the DC motor and the motor starts rotating in the desired direction. As the motor shaft progresses, the potentiometer knob also rotates as it is coupled with a motor shaft with the help of a gear arrangement. As the position of the potentiometer knob changes, there is an electrical signal produced at the potentiometer port. As the angular position of the potentiometer knob progresses, the output or feedback signal increases. After the

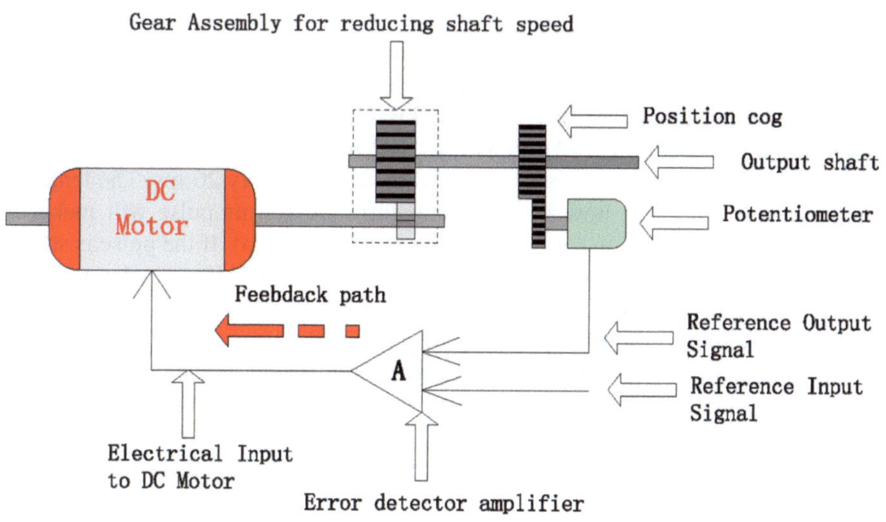

Fig. 4.14 Schematic of DC servo motor

desired angular position of the motor shaft is achieved, the potentiometer knob reaches to a position such that the electrical signal generated in the potentiometer becomes the same as that of the external electrical signal given to the amplifier. At this stage, there is no output signal from the amplifier to the motor input, as there is no difference between the externally applied signal and the signal generated at the potentiometer. As a result, the motor stops rotating.

Servo motors are being used in a wide variety of robots, machines, and general robotic applications, including: robotic arms, radio-controlled toy cars, airplanes, and helicopters, industrial machinery, and many more applications. There are many reasons that make servo motors so common—their ease of control, the low energy requirements (efficiency), the high torque, TTL voltage level control, and even the physical properties; servo motors are relatively small sized and have a low weight.

FYI	Comparison between servo motors and stepper motors ■ The most significant difference between servo motors and stepper motors is the fact that servo motors operate in a closed loop, while stepper motors operate in an open loop. This means that servo motors have an internal feedback—they are able to measure their position, the difference between the actual position and the desired position, and to fix the gap by controlling the motor. Stepper motors, on the other hand, have no feedback and thus are more error-prone ■ RC Servo motors are limited to 0°–180° of movement and require physical and electrical modification in order to be able to move in 360°. Stepper motors do not have this limit ■ Stepper motors are usually cheaper than servo motors ■ Stepper motors lose torque in high rotational speeds, while servo motors do not

4.3.2 Driven Circuit Design

Servo motors are controlled through a control line, usually yellow in color. The pulse width of the signal sent to the servo control wire determines how the motor will move, either clockwise or counterclockwise. The signal is known as a pulse proportional modulation (PPM). PPM uses 1–2 ms out of a 20 ms time period to encode its information. The servo expects to see a pulse every 20 ms. The length of the pulse will determine how far the motor turns. A 1.5 ms pulse will make the motor turn to a 90° position (often called the neutral position). If the pulse is shorter than 1.5 ms, then the motor will turn the shaft closer to 0°. If the pulse is longer than 1.5 ms, the shaft turns closer to 180°. Figure 4.15 shows how different pulse widths correspond to different positions of the motor.

The amount of power applied to the motor is proportional to the distance it needs to travel. Therefore, if the shaft needs to turn a large distance, the motor will run at full speed. If it needs to turn only a small amount, the motor will run at a slower speed. When the servo motor reaches the desired position, it will hold there until a signal is sent to move.

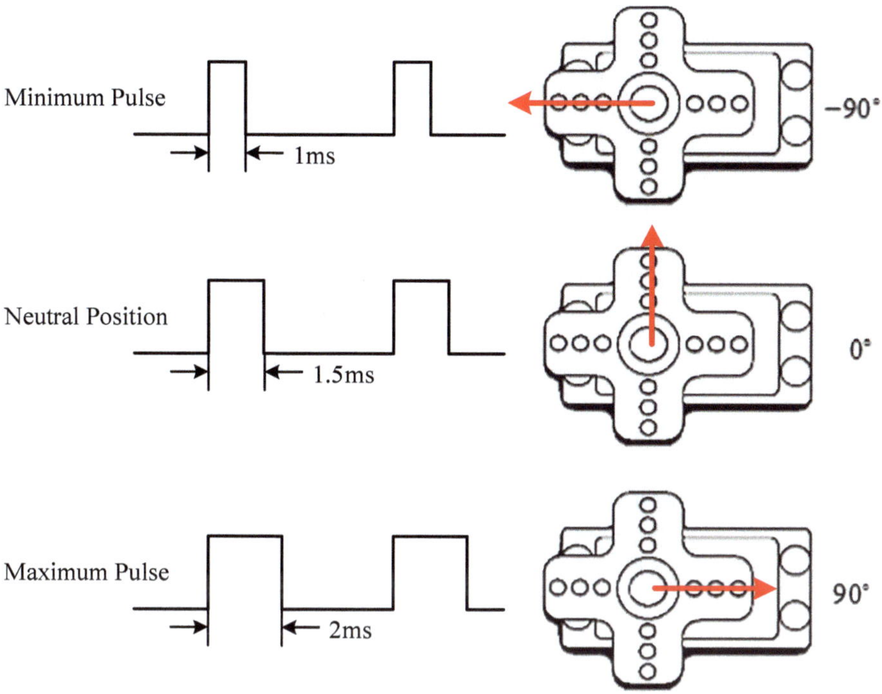

Fig. 4.15 Pulse widths and their corresponding position of the servo motor

4.3.3 *Demonstration*

1. **Components**

- DFRobot UNO R3 board and USB cable × 1
- 10 k ohm potentiometer × 1
- Servo motor × 1
- Jumper wires × *n*

4.4 Hardware Setting

Servo motors have three wires: power, ground, and signal. The power wire is typically red, and should be connected to the 5 V pin on the UNO board. The ground wire is typically black or brown and should be connected to a ground pin on the board. The signal pin is typically yellow or orange and should be connected to Pin 9 on the board. (shown in Fig. 4.16). Furthermore, a potentiometer is also used to control the servo position.

1. **Sample Codes**

```
#include <Servo.h>
Servo myservo; // create servo object to control a servo
int potpin = 0; // analog pin used to connect the potentiometer
int val; // variable to read the value from the analog pin
void setup() {
  myservo.attach(9); // attaches the servo on pin 9 to the servo
object
}

void loop() {
  val = analogRead(potpin); // reads the value of the potentiometer
(value between 0 and 1023)
  val = map(val, 0, 1023, 0, 180); // scale it to use it with the
servo (value between 0 and 180)
  myservo.write(val); // sets the servo position according to the
scaled  value
  delay(15); // waits for the servo to get there
}
```

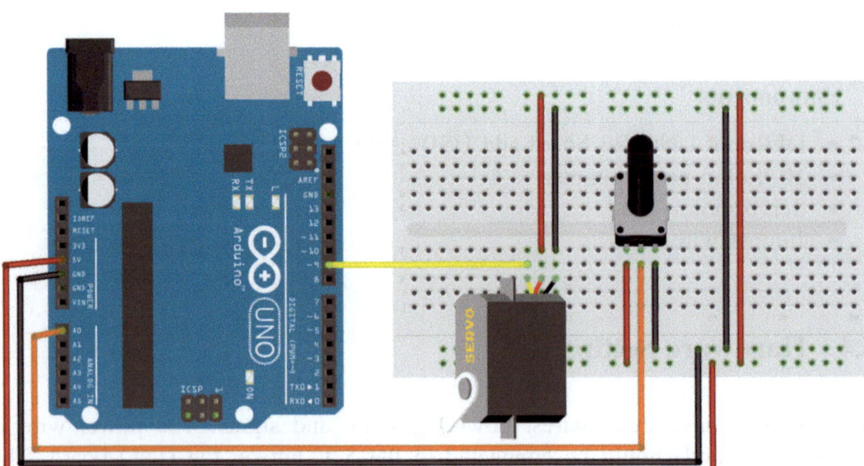

Fig. 4.16 Servo motor and Romeo board connection diagram

4.5 Explanation

#include < Servo.h>

The #include <Servo.h> loads the necessary code to use a servo motor. After including it, you can initialize the servo motor with Servo myservo. which defines the motor and gives it the name myservo, which is how it will be referred to in the rest of the code. Before using it, you should download the library from http://playground.arduino.cc/uploads/ComponentLib/SoftwareServo.zip.

myservo.attach(pin)

myservo.attach(9) turns on the servo and informs the code that the servo can be accessed through digital Pin 9.

There is another syntax of attach(): **servo.attach(pin, min, max)**

pin: the number of the pin that the servo is attached to

min (optional): the pulse width, in microseconds, corresponding to the minimum (0°) angle on the servo (defaults to 544)

max (optional): the pulse width, in microseconds, corresponding to the maximum (180°) angle on the servo (defaults to 2400)

myservo.write(angle)

This code will inform the servo motor to move forward for the degree within the servo1.write(val) command, then waits 15 ms in the delay(15)command

angle: the value to write to the servo, from 0 to 180.

myservo.read()

Reads the current angle of the servo (the value passed to the last call to write()). The angle of the servo, from 0° to 180°.

myservo.detach()

Detach the myservo variable from its pin. If all myservo variables are detached, then Pins 9 and 10 can be used for PWM outputs with analogWrite().

myservo.writeMicroseconds(uS)

Writes a value in microseconds (uS) to the servo, controlling the shaft accordingly. On a standard servo, this will set the angle of the shaft. On standard servos, a parameter value of 1000 is fully counterclockwise, 2000 is fully clockwise, and 1500 is in the middle.

Note that some manufactures do not follow this standard very closely so that servos often respond to values between 700 and 2300. Feel free to increase these endpoints until the servo no longer continues to increase its range. Note however that attempting to drive a servo past its endpoints (often indicated by a growling sound) is a high-current state, and should be avoided.

myservo.readMicroseconds()

uS: the value of the parameter in microseconds (*int*)

Returns the last written servo pulse width in microseconds.

Chapter 5
Wireless Control Using the Arduino

5.1 Infrared Transmitter and Receiver Module

5.1.1 Introduction

An infrared (IR) transmitter consists of an infrared LED array, which functions as a luminophore, and a PN junction made of a special material with high infrared radiation efficiency, which is usually GaAs. When a current is injected into the PN junction by a forward bias voltage, it can excite a source of infrared light with a center wavelength range of 830–950 nm. The power of the infrared light excited is proportional to the current injected. However, in the case that the injected current exceeds the maximum rating, the power of the infrared light may decline as the current increases.

The infrared receiver is a semiconductor device used for translating infrared light signals into electrical signals. The core component of it is a PN junction made of a special material. The PN junction has a different structure from a general purpose diode, enabling more infrared light to be received. As the intensity of the infrared light increases, more current can be generated.

The schematic diagram of an IR transmitter/receiver is shown in Fig. 5.1.

In an IR transmitter/receiver module, the data being transmitted is wrapped by start and stop bits. The start bits alert the receiver that a command will soon follow, and the stop bits indicate that the transmission has ended. One downfall of using infrared for data transmission is the presence of infrared sources other than that of the data transmitter. Because of environmental infrared noise, infrared transmitters are designed to oscillate at a certain frequency. The receiver is simply "tuned" to react to that frequency. It will ignore any infrared signals that are outside of that carrier frequency, much the way your radio is tuned to your favorite FM channel.

T. Pan and Y. Zhu, *Designing Embedded Systems with Arduino*,
DOI 10.1007/978-981-10-4418-2_5

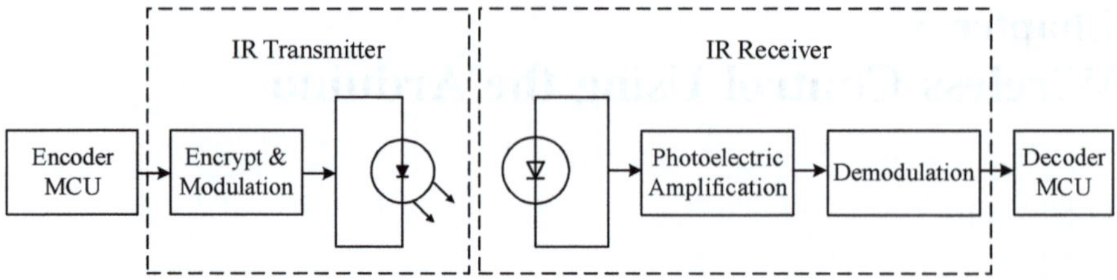

Fig. 5.1 IR transmitter/receiver principle

A common modulation scheme for IR communication is somewhere around 38 kHz. There are very few natural sources that have the regularity of a 38 kHz signal, so an IR transmitter sending data at that frequency would stand out among ambient IR. 38 kHz modulated IR data is most common, but other frequencies are also used.

5.1.2 IR Transmitter/Receiver Module

1. **Components**
   - DFRobot UNO R3 board and USB cable × 2.
   - IR Receiver Module (SKU:DFR0094) × 1.
   - IR Transmitter Module (SKU:DFR0095) × 1.
   - Jumper wires × *n*.
2. **Hardware setting** (Figs. 5.2 and 5.3).

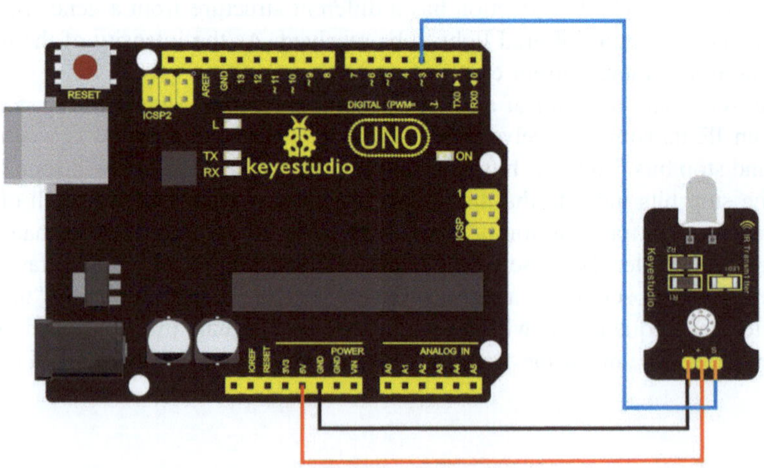

Fig. 5.2 A diagram of the layout of the IR transmitter and UNO R3

Fig. 5.3 A diagram of the layout of the IR receiver and UNO R3

3. **Sample Codes**
   The first thing you need to do is install the IR Arduino library, the IR library can be found at https://github.com/z3t0/Arduino-IRremote.
   The codes for the IR Transmitter:

```
#include <IRremote.h>
IRsend irsend;
void setup() {
}

void loop() {
  irsend.sendRC5(0x0, 8); //send 0x0 code (8 bits)
  delay(200);
  irsend.sendRC5(0x1, 8);
  delay(200);
}
```

The codes for the IR receiver:

```
#include <IRremote.h>
const int RECV_Pin = 11; // IR Sensor pin
const int LED_Pin = 13; // LED pin

IRrecv irrecv(RECV_Pin);
decode_results results;

void setup() {
  Serial.begin(9600); // configure the baudrate
  irrecv.enableIRIn(); // Start the receiver
}

void loop() {
  if (irrecv.decode(&results)) {
    if (results.bits > 0) {
      int state;
      if (0x1 == results.value) {
        state = HIGH;
      }
      else {
        state = LOW;
      }
      digitalWrite(LED_Pin, state);
    }
    irrecv.resume(); // prepare to receive the next value
  }
}
```

4. **Results**
   The LED of the shield connected to the IR Receiver will blink when the IR Receiver faces the IR Transmitter.

5.1.3 IR Kit

1. **Comp7onents**
   - DFRobot UNO R3 microcontroller board and USB cable × 1.
   - IR Kit (SKU:DFR0107) × 1.
   - Jumper wires × *n*.
2. **Hardware setting** (Fig. 5.4)

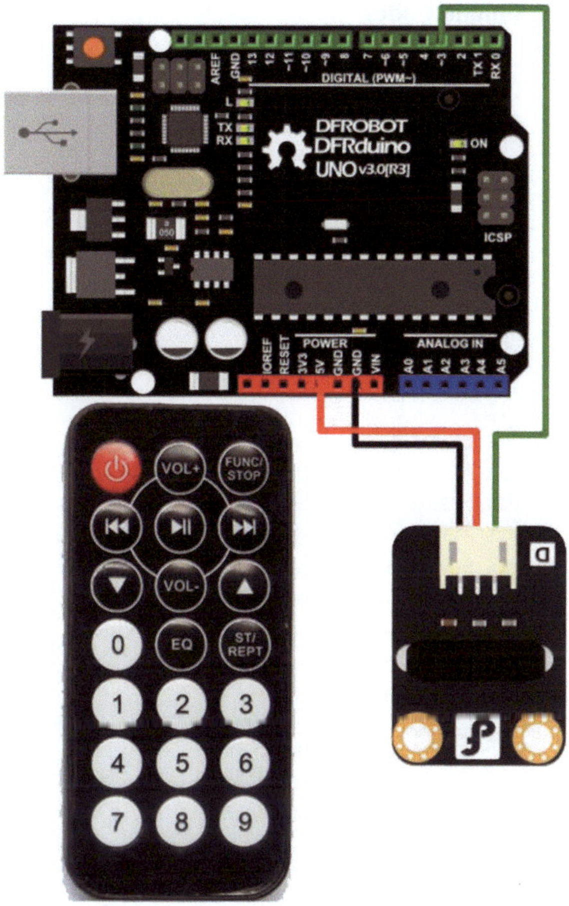

Fig. 5.4 A diagram of the layout of DFR0107 and UNO R3

3. Sample Codes

```
#define IR_BIT_LENGTH 32 // number of bits sent by IR remote
#define FirstLastBit 15 // divide 32 bits into two 15 bit chunks
for integer variables. Ignore center two bits. they are all the
same.
#define BIT_1 1500  // Binary 1 threshold (Microseconds)
#define BIT_0 450  // Binary 0 threshold (Microseconds)
#define BIT_START 4000 // Start bit threshold (Microseconds)

#define IR_Pin 8  // IR Sensor pin
#define LED_Pin 13  // LED goes off when signal is received

int debug = 0;  // flag as 1 to output raw IR pulse data stream
length in microseconds
int output_verify = 0;  // flag as 1 to print decoded verification
integers. same number for all buttons
int output_key = 0;  // flag as 1 to print decoded key integers
int remote_verify = 16128;  // verifies first bits are
11111100000000 different remotes may have different start codes

void setup() {
  pinMode(LED_Pin, OUTPUT); //configure the pin
  pinMode(IR_Pin, INPUT);
  digitalWrite(LED_Pin, LOW); // turn off
  Serial.begin(9600);
}

void loop() {
  digitalWrite(LED_Pin, HIGH);
  int key = get_ir_key();
  digitalWrite(LED_Pin, LOW);  // turn LED off while processing
response
  do_response(key);
  delay(130);  // 2 cycle delay to cancel duplicate keypresses
}

/*  wait for a keypress from the IR remote, and return the
  integer mapping of that key (e.g. power button on remote returns
the integer 1429) */
int get_ir_key()  {
  int pulse[IR_BIT_LENGTH];
  int bits[IR_BIT_LENGTH];

  do {} //Wait for a start bit
```

```
  while (pulseIn(IR_Pin, HIGH) < BIT_START);

  read_pulse(pulse);
  pulse_to_bits(pulse, bits);
  RemoteVerify(bits);
  return bits_to_int(bits);
}

/* use pulseIn to receive IR pulses from the remote.
  Record the length of these pulses (in ms) in an array */
void read_pulse(int pulse[]) {
  for (int i = 0; i < IR_BIT_LENGTH; i++) {
    pulse[i] = pulseIn(IR_Pin, HIGH);
  }
}

/* IR pulses encode binary "0" as a short pulse, and binary "1"
as a long pulse.  Given an array containing pulse lengths,
  convert this to an array containing binary values */
void pulse_to_bits(int pulse[], int bits[]) {
  if (debug) {
    Serial.println("-----");
  }
  for (int i = 0; i < IR_BIT_LENGTH; i++) {
    if (debug) {
      Serial.println(pulse[i]);
    }
    if (pulse[i] > BIT_1) { //is it a 1?
      bits[i] = 1;
    }
    else if (pulse[i] > BIT_0) { //is it a 0?
      bits[i] = 0;
    }
    else { //data is invalid...
      Serial.println("Error");
    }
  }
}

/* check returns proper first 14 check bits */
void RemoteVerify(int bits[]) {
  int result = 0;
  int seed = 1;

```

```
  //Convert bits to integer
  for (int i = 0 ; i < (FirstLastBit) ; i++) {
    if (bits[i] == 1) {
      result += seed;
    }
    seed *= 2;
  }
  if (output_verify) {
    Serial.print("Remote ");
    Serial.print(result);
    Serial.println(" verification code");
  }
  if (remote_verify != result) {
    delay (60);  //verify first group of bits. delay for data
stream to end, then try again.
    get_ir_key();
  }
}

/* convert an array of binary values to a single base-10 integer
*/
int bits_to_int(int bits[]) {
  int result = 0;
  int seed = 1;

  //Convert bits to integer
  for (int i = (IR_BIT_LENGTH - FirstLastBit) ; i < IR_BIT_LENGTH ;
i++) {
    if (bits[i] == 1) {
      result += seed;
    }
    seed *= 2;
  }
  return result;
}

/* respond to specific remote-control keys with different
behaviors */
void do_response(int key) {
  if (output_key) {
    Serial.print("Key ");
    Serial.println(key);
  }

```

```
    switch (key) {
      case 32640:  // turns on UUT power
        Serial.println("POWER");
        break;
      case 32385:  // FUNC/STOP turns off UUT power
        Serial.println("FUNC/STOP");
        break;
      case 32130:  // |<< ReTest failed Test
        Serial.println("|<<");
        break;
      case 32002:  // >|| Test
        Serial.println(">||");
        break;
      case 31875:  // >>| perform selected test number
        Serial.println(">>|");
        break;
      case 32512:  // VOL+ turns on individual test beeper
        Serial.println("VOL+");
        break;
      case 31492:  // VOL- turns off individual test beeper
        Serial.println("VOL-");
        break;
      case 31620:  // v scroll down tests
        Serial.println("v");
        break;
      case 31365:  // ^ scroll up tests
        Serial.println("^");
        break;
      case 30982:  // EQ negative tests internal setup
        Serial.println("EQ");
        break;
      case 30855:  // ST/REPT Positive tests Select Test and Repeat
Test
        Serial.println("ST/REPT");
        break;
      case 31110:  // 0
        Serial.println("0");
        break;
      case 30600:  // 1
        Serial.println("1");
        break;
      case 30472:  // 2
        Serial.println("2");
        break;
```

```
    case 30345:  // 3
      Serial.println("3");
      break;
    case 30090:  // 4
      Serial.println("4");
      break;
    case 29962:  // 5
      Serial.println("5");
      break;
    case 29835:  // 6
      Serial.println("6");
      break;
    case 29580:  // 7
      Serial.println("7");
      break;
    case 29452:  // 8
      Serial.println("8");
      break;
    case 29325:  // 9
      Serial.println("9");
      break;
    default: {
        Serial.print("Key ");
        Serial.print(key);
        Serial.println(" not programmed");
      }
      break;
  }
}
```

4. **Results**
   The sketch will automatically decode the type of remote you are using and identify which button on your remote has been pressed. Open the serial port in the Arduino IDE at 9600 bps and hit different buttons on your remote (shown in Fig. 5.3) (Fig. 5.5).
   In IR kit, every button on the remote control has a corresponding value (shown in Table 5.1).

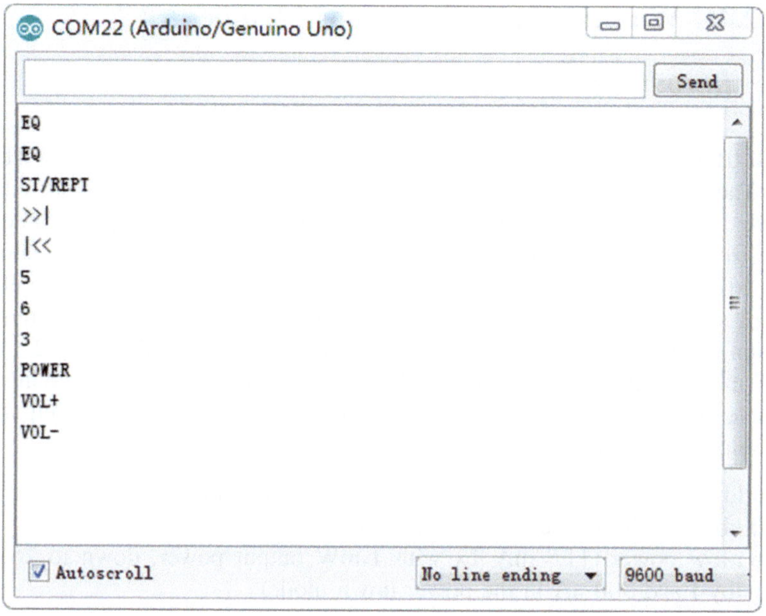

Fig. 5.5 DFR0107 output in Arduino serial monitor

Table 5.1 The value of characters in remote control

Characters in remote control	Value	Characters in remote control	Value
Power (red)	0xff00	ST/REPT	0xf10e
VOL+	0xfe01	1	0xef10
FUNC/STOP	0xfd02	2	0xee11
⏮	0xfb04	3	0xed12
⏯	0xfa05	4	0xeb14
⏭	0xf906	5	0xea15
▼	0xf708	6	0xe916
VOL−	0xf609	7	0xe718
▲	0xf50a	8	0xe619
0	0xf30c	9	0xe51a
EQ	0xf20d		

5.2 2.4G Wireless Radio Frequency Module

5.2.1 Introduction

nRF24L01 is a single chip radio transceiver for the worldwide 2.4–2.5 GHz ISM band. The channel spacing is 1 MHz, which allows for 125 possible channels numbered 0, 1, 2, …, 124. The transceiver consists of a fully integrated frequency synthesizer, a power amplifier, a crystal oscillator, a demodulator, modulator, and enhanced ShockBurst™ protocol engine. Output power, frequency channels, and protocol setup are easily programmable through an SPI interface. Current consumption is very low, only 9.0 mA at an output power of −6 dBm and 12.3 mA in the RX mode. Built in Power Down and Standby modes makes power saving easily realizable. The key features of the nRF24L01 are as follows (source, http://www.nordicsemi.com/eng/Products/2.4 GHz-RF/nRF24L01P):

- Worldwide 2.4 GHz ISM band (free, unlicensed band)
- 250 kbps, 1 Mbps, and 2 Mbps on air data rates
- Ultra-low power (11.3 mA Tx with 1 mW output power, down to 26 μA in standby-I and 900 nA in the power down mode)
- 1.9–3.6 V supply voltage, with 5 V tolerant input pins
- Automatic acknowledgment message transmissions with automatic retries
- RX and TX FIFO's with ACK user data possibility
- Up to 6 data pipes/addresses for a simplified star network

5.2.2 2.4 GHz Wireless RF Transceiver Module

NRF24L01 wireless RF transceiver module is a great wireless module suitable for short range 100 m remote controls at a 250 kbps data rate. They are transceivers, which means that each module can transmit and receive data. The module shape is shown in Fig. 5.2 and has the following pins connected to a microcontroller (Fig. 5.6):

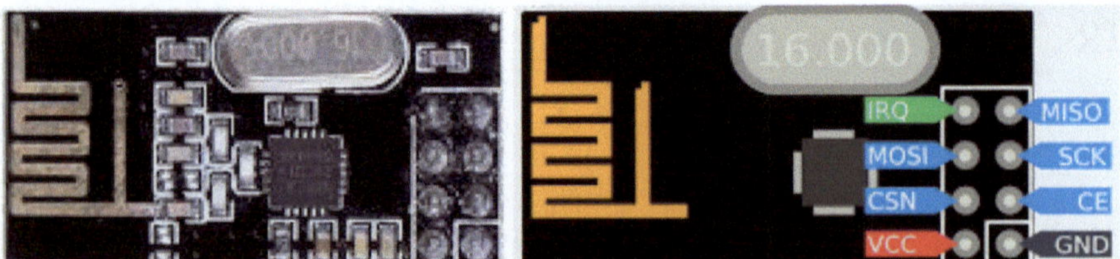

Fig. 5.6 nRF24L01 Wireless RF Transceiver Module

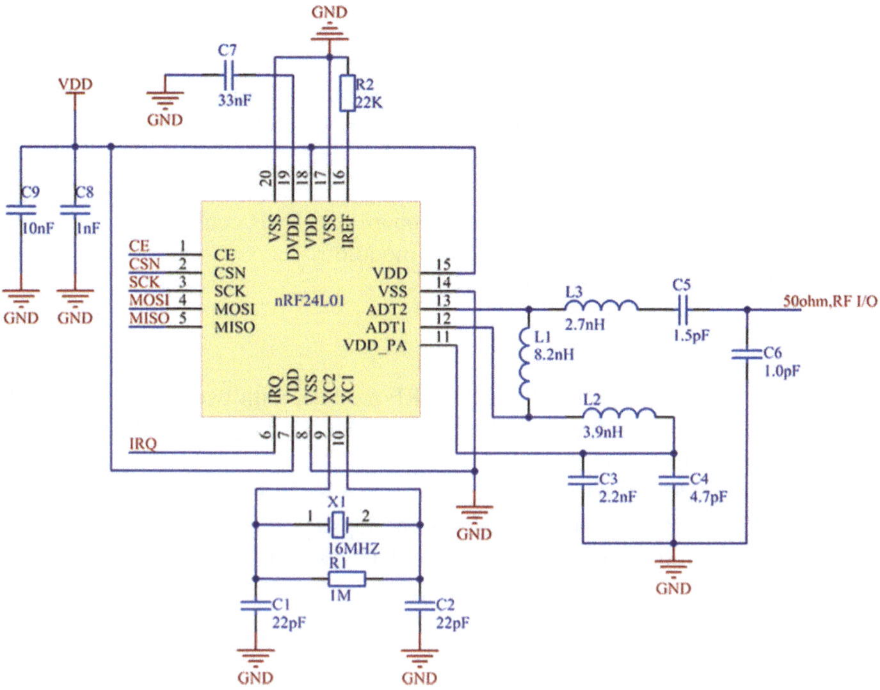

Fig. 5.7 Schematic of NRF24L01 wireless RF transceiver module

- **GND**: Ground.
- **VCC**: 3.3 V.
- **CE**: Chip (RX/TX) Enable, high active. If high, module is either sending or listening.
- **CSN**: Chip Select Not, low active. If low, the chip responds to SPI commands. This is actually the "real" chip select signal, and is easily confused with CE, which enables/disables the transceiver radio part.
- **SCK**: SPI Shift Clock, up to 10 MHz.
- **MOSI**: Master-Out-Slave-In, used to shift data from the microcontroller to the device.
- **MISO**: Master-In-Slave-Out, used to shift data from the device to the microcontroller.
- **IRQ**: Optional Interrupt Request pin. Signals RX/TX status such as packet sent or received.

The schematic diagram of the module is shown in Fig. 5.7:

5.2.3 Demonstration

The experiment is to have the A0 and A1 analog pins collect the X and Y values of the joystick, and send the data wirelessly via the NRF24L01 module.

1. **Components**
   - DFRobot UNO R3 microcontroller board and USB cable × 2.
   - NRF24L01 wireless RF transceiver module × 2.
   - DFR0061 (joystick module) × 1.
   - Jumper wires × *n*.
2. **Hardware setting**
   There are eight pins on the NRF24L01 RF module, with two power pins for the VCC and GND, the CE pin, SCN pin, SCK, MOSI, MISO, and IRQ pin. Refer below for the hardware setup.
   To setup the NRF24L01 as a transmitter on the Arduino, the wiring connections are shown in Fig. 5.8. In this diagram, the joystick module is powered by 5 V and GND on Arduino, while Horizontal (*X* axis) is set to A0 and Vertical (*Y* axis) is set to A1 of Arduino, we leave the Select (*Z* axis for "1" and "0") unconnected.
   To setup the NRF24L01 RF module as a receiver to sync the data at 2.4 GHz band, we setup a separate Arduino and NRF24L01 module as in the setup in the picture below. The wire connections are the same as the transmitter (Table 5.2; Fig. 5.9).

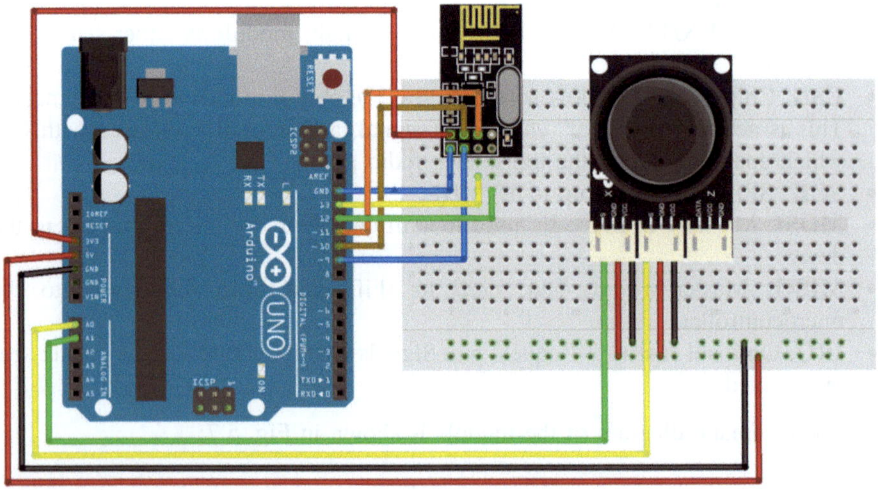

Fig. 5.8 A transmitter diagram of the layout of the nRF24L01 and UNO R3

Table 5.2 Wiring connecting between Arduino UNO and nRF24L01

Number	nRF24L01	Arduino UNO
1	VCC	3.3 V
2	GND	GND
3	CE	D9
4	CSN	D10
5	SCK	D13
6	MOSI	D11
7	MISO	D12
8	IRQ	Un-connecting

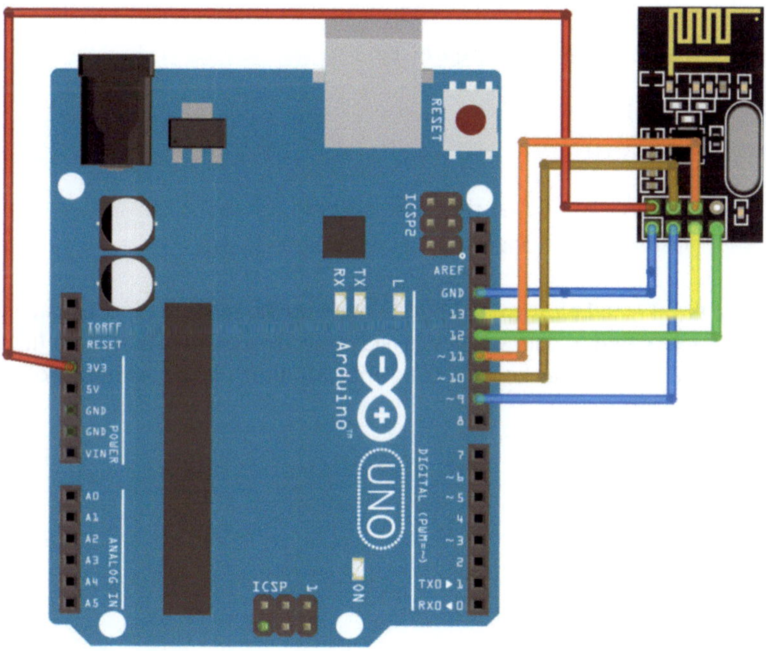

Fig. 5.9 A Receiver diagram of the layout of the nRF24L01 and UNO R3

3. **Sample Codes**
   Download the RF24 Library (https://github.com/maniacbug/RF24) for NRF24L01, and write the sketch below to be uploaded to Arduino.

```
#include <SPI.h>
#include <nRF24L01.h>
#include <RF24.h>
#define CE_PIN 9
#define CSN_PIN 10
#define JOYSTICK_X A0
#define JOYSTICK_Y A1

const uint64_t pipe = 0xE8E8F0F0E1LL;
RF24 radio(CE_PIN, CSN_PIN);
int joystick[2];
void setup() {
  Serial.begin(9600);
  radio.begin();
  radio.openWritingPipe(pipe);
}

void loop() {
  joystick[0] = analogRead(JOYSTICK_X);
  joystick[1] = analogRead(JOYSTICK_Y);
  radio.write( joystick, sizeof(joystick) );
}
```

Upload the sketch below to Arduino to open up the reading pipe, listen, and read the joystick X and Y data wirelessly via radio frequency. The data is then serial printed and can be checked via the Serial Monitor on the Arduino software.

```
#include <SPI.h>
#include <nRF24L01.h>
#include <RF24.h>
#define CE_PIN 9
#define CSN_PIN 10

const uint64_t pipe = 0xE8E8F0F0E1LL;
RF24 radio(CE_PIN, CSN_PIN);
int joystick[2];

void setup() {
  Serial.begin(9600);
  delay(1000);
  Serial.println(“Nrf24L01 Receiver Starting”);
  radio.begin();
  radio.openReadingPipe(1, pipe);
  radio.startListening();;
}

void loop() {
  if ( radio.available() ) {
    bool done = false;
    while (!done) {
      done = radio.read( joystick, sizeof(joystick) );
      Serial.print(“X = ”);
      Serial.print(joystick[0]);
      Serial.print(“Y = ”);
      Serial.println(joystick[1]);
    }
  }
  else {
    Serial.println(“No radio available”);
  }
}
```

4. **Results**
   Connect the receiver to the computer, open up the Arduino software and Serial Monitor, the X and Y data of the joystick shall be printed out as in Fig. 5.10.

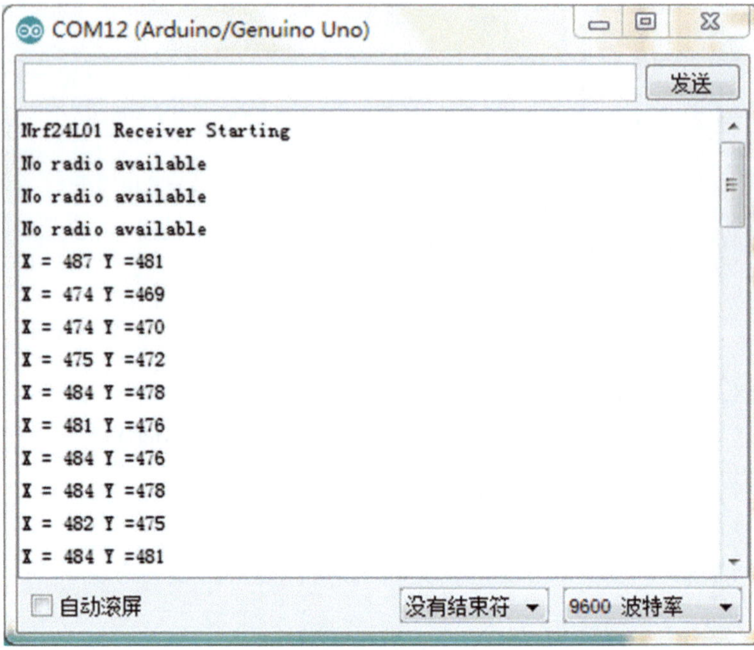

Fig. 5.10 nRF24L01 result in Arduino serial monitor

5.3 Bluetooth Module

5.3.1 Introduction

Bluetooth® wireless technology is becoming a popular standard in communication. It is one of the fastest growing fields in wireless technologies. It is convenient to use and has the bandwidth to meet most of today's demands for mobile and personal communications (using short-wavelength UHF radio waves in the ISM band from 2.4 to 2.485 GHz). Bluetooth technology handles the wireless part of the communication channel; it transmits and receives data wirelessly between these devices. It delivers the received data and receives the data to be transmitted to and from a host system through a host controller interface (HCI). The most popular host controller interface today is either a UART or a USB.

Bluetooth is a packet-based protocol with a master–slave structure. One master may communicate with up to seven slaves in a piconet. All devices share the master's clock. Packet exchange is based on the basic clock, defined by the master, which ticks at 312.5 μs intervals. Two clock ticks make up a slot of 625 μs, and two slots make up a slot pair of 1250 μs. In the simple case of single-slot packets, the master transmits in even slots and receives in odd slots. The slave, conversely, receives in even slots and transmits in odd slots. Packets may be 1, 3, or 5 slots long, but in all cases the master's transmission begins in even slots and the slave's in odd slots.

5.3.2 HC-05 Module

The HC-05 module is one of the easiest Bluetooth module that we can find in the market, it is also very cheap and suitable for those who are cost conscious. The module shape is shown in Fig. 5.11 and has the following pins connected to a microcontroller:

- **EN/KEY**: If the pin is set as HIGH before the power is applied, it forces the AT Command Setup Mode. LED blinks slowly (2 s)
- **VCC**: +5 Power
- **GND**: System/ Arduino Ground
- **TX**: Transmit Serial Data from HC-05 to Arduino Serial Receive. NOTE: 3.3 V HIGH level: OK for Arduino
- **RX**: Receive Serial Data from Arduino Serial Transmit
- **STATE**: Informs if connected or not

The module has two modes of operation, Command Mode where we can send AT commands to it and Data Mode where it transmits and receives data to another Bluetooth module. The default mode is the DATA Mode, and this is the default configuration, which may work fine for many applications:

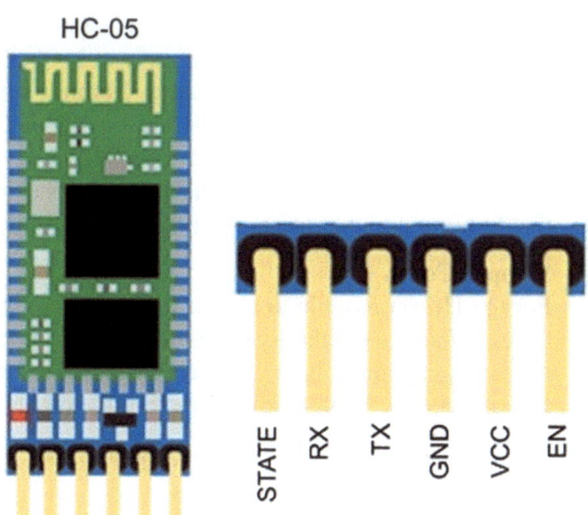

Fig. 5.11 HC-05 Module

- Baud Rate: 9600 bps, Data: 8 bits, Stop Bits: 1 bit, Parity: None, Handshake: None
- Passkey: 1234

In some cases, you may want to change some of the configuration setup values. HC-05 comes with a rich set of AT commands (shown in Table 5.3) to perform various tasks such as changing the module's default settings including changing the pass code, the device name, and the baud rate.

The HC-05 module can switch roles from the master to slave structure; its original password is 1234. As a master structure, the HC-05 module cannot store a slave structure in its original setting, but only after communicating the slave structure by setting the AT command—AT + CMODE = 0; it can, however, communicate with any slave. It can also communicate by appointing specific addresses of mobile phones, computers, and slave structures. In the meantime, the master structure is able to automatically scan slave structures by default.

5.3.3 Modify HC-05 Module Defaults Using at Commands

1. **Components**
   - DFRobot UNO R3 board and USB cable × 1.
   - HC-05 Bluetooth Module × 1.
   - Jumper wires × *n*.
2. **Hardware setting**
   In this experiment, the Arduino does two things. It takes the AT commands you enter from the Arduino IDE Serial Monitor and sends those commands to HC-05. The program then reads the output of HC-05 and displays it on the Arduino IDE Serial Monitor.
   The Arduino communicates with the HC-05 using SoftwareSerial ports while the Arduino communicates with the user via the Serial Monitor (Fig. 5.12).
3. **Sample Codes**
   For the HC-05 module to switch to the AT command mode, the HC-05 pin 34 (often referred to as the Key pin) needs to be pulled HIGH but in a certain order of events explained below. When in the AT command mode, the LED on the HC-05 needs to blink on/off every second and the HC-05 needs to communicate at 38,400 baud rate.

Table 5.3 The main AT commands of bluetooth

Command	Response	Parameter	Description
AT	OK	None	Test
AT + RESET	OK	None	Reset
AT + VERSION?	+VERSION: <Param> OK	Version number	Get the soft version
AT + ORGL	OK	None	Restore default status
AT + ADDR?	+ADDR: <Param> OK	Bluetooth address	Get module bluetooth address
AT + NAME?	+NAME <Para1> OK	Bluetooth device name Default: "HC-05"	Set/inquire device's name
AT + NAME = <Para1>	1. +AT + NAME: <Para1> OK (success) 2. ERROR = <Error_Code> (failure)		
AT + ROLE?	+ROLE: <Param> OK	0-Slave 1-Master 2-Slave-Loop Default: 0	Set/inquire module role (master of slave)
AT + ROLE = <Param>	OK		
AT + PSWD = <Param>	OK	Param: passkey Default: "1234"	Set/inquire-passkey
AT + PSWD?	+ PSWD: <Param> OK		
AT + UART?	+UART = <Param1>, <Param2>, <Param3> OK	Para1: baud rate (bits/s), the value (Decimal) should be one of the following: 4800, 9600, 19,200, 38,400, 57,600, 115,200, 23,400, 460,800, 921,600, 1,382,400 Para2: stop bit 0–1 bit 1–2 bits Para3: parity bit	Set/Inquire-serial parameter
AT + UART = <Para1>, <Para2>, <Para3>	OK		
AT + STATE?	+STATE: <Param> OK	"INITIALIZED"-initialized status "READY"-ready status	Get the work status of bluetooth module

(continued)

Table 5.3 (continued)

Command	Response	Parameter	Description
		"PAIRABLE"-pairable status "PAIRED"-paired status "INQUIRING"-inquiring status "CONNECTING-connecting status "CONNECTED"-connected status "DISCONNECTED"-disconnected status "NUKNOW"-unknown status	
AT + CMODE = 1	OK	None	Make the master module pair with the slave module

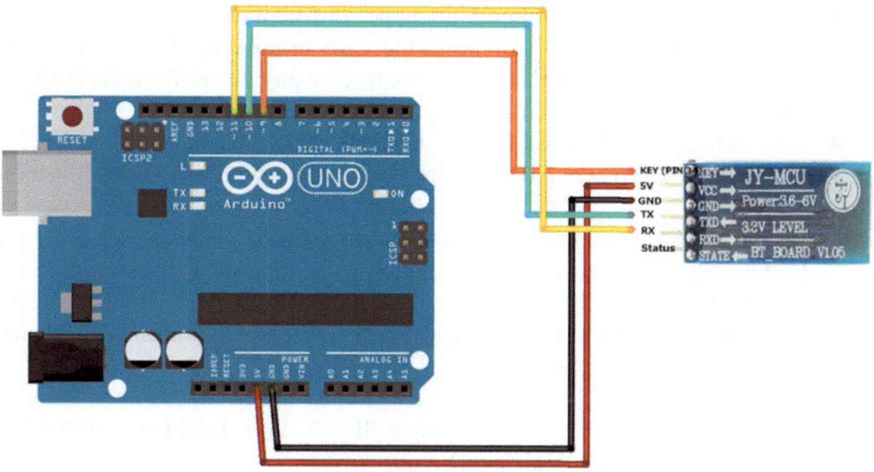

Fig. 5.12 A diagram of the layout of the HC-05 and UNO R3

```
#include <SoftwareSerial.h>

SoftwareSerial BTSerial(10, 11); // RX | TX

void setup() {
  pinMode(9, OUTPUT);  // this pin will pull the HC-05 pin 34 (key
pin) HIGH to switch module to AT mode
  digitalWrite(9, HIGH);
  Serial.begin(9600);
  Serial.println("Enter AT commands:");
  BTSerial.begin(38400); // HC-05 default speed in AT command
}

void loop() {
// Keep reading from HC-05 and send to Arduino Serial Monitor
  if (BTSerial.available())
    Serial.write(BTSerial.read());

// Keep reading from Arduino Serial Monitor and send to HC-05
  if (Serial.available())
    BTSerial.write(Serial.read());
}
```

Follow these steps in the stated order to switch the HC-05 to the AT command mode.

- Wire the HC-05 and Arduino Uno as per instructions.
- Before you connect the Arduino to the USB remove the VCC (power) red wire from the HC-05 so that it does not receive any power from the Arduino. All other wires are still connected.
- Now connect the Arduino Uno to the USB cable connected to your PC.
- Make sure the HC-05 module is NOT PAIRED with any other Bluetooth device.
- Reconnect the Arduino Uno 5 V wire to the HC-05's VCC (5 V power) pin. In this step, you need to hold the button switch closed while powering on.
- The HC-05 LED will blink on and off at about 2 s intervals. Now the HC-05 is in the AT command mode and ready to accept commands to change configuration and settings.
- To test if everything is wired correctly, open the Serial Monitor from the Arduino IDE and type "AT" and click SEND. You should see an "OK".
- If you do not see an "OK" check your wiring.

4. **Results**
   Once you open the serial monitor and arrange line settings to NL & CR and baud rate to 38,400 you will manage to communicate with the module. Type "AT" and the module should respond as "OK" (Fig. 5.13).

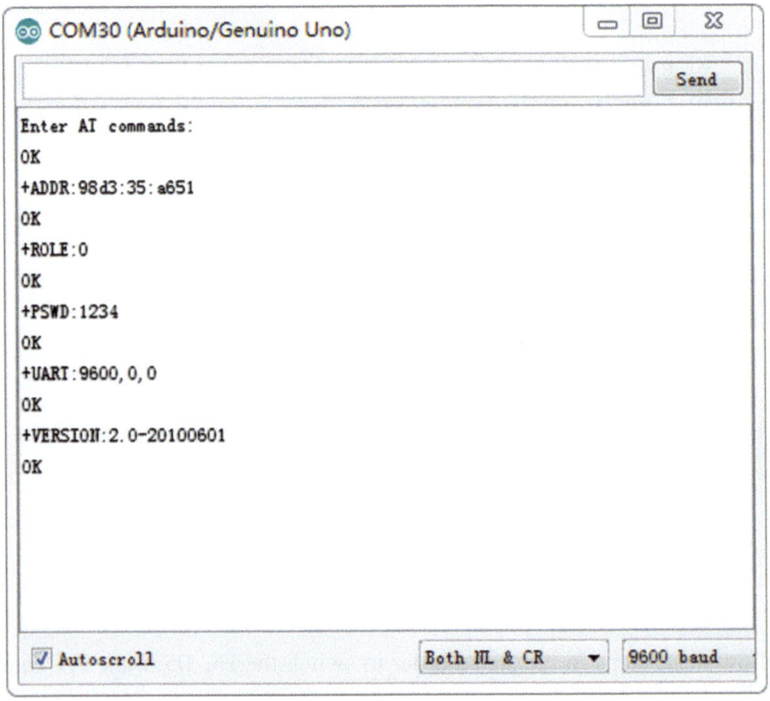

Fig. 5.13 HC-05 result in Arduino serial monitor

If your module is not responding, try to change the baud rate to 9600 and check the wiring again.

5.3.4 Demonstration

In this experiment, you will learn how to make a connection between two HC-05 modules.

1. **Components**
   - DFRobot UNO R3 board and USB cable × 2.
   - HC-05 Bluetooth Module × 2.
   - Jumper wires × *n*.
2. **Hardware setting**
   Both Bluetooth modules are connected in the same way (see Table 5.4). The SoftwareSerial ports is used to communicate between the user and Arduino via the Serial Monitor. The circuits is shown as follows (Fig. 5.14):

Table 5.4 Wiring connecting between Arduino UNO and nRF24L01

Number	HC-05	Arduino UNO
1	VCC	5 V
2	GND	GND
3	TX	D10
4	RX	D11
5	KEY	D9

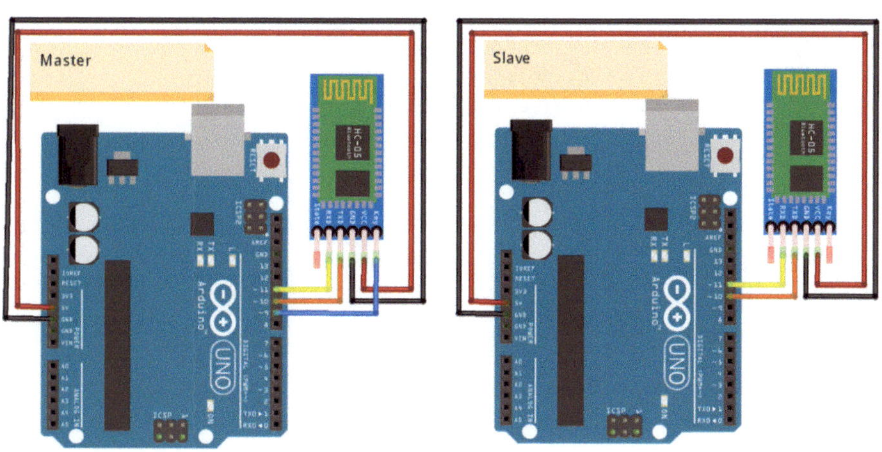

Fig. 5.14 The diagram of layout of two HC-05 modules

3. **Sample Codes**

(1) Codes for Master

```
// Connect the HC-05 and communicate using the serial monitor
// When first powered on, you need to manually enter AT mode
// The default baud rate for AT mode is 38400
#include <SoftwareSerial.h>

SoftwareSerial BTSerial(10, 11); // RX | TX
// Connect the HC-05 TX to Arduino pin 10
// Connect the HC-05 RX to Arduino pin 11

void setup() {
  pinMode(9, OUTPUT);  // this pin will pull the HC-05 pin 34 (key
pin) HIGH to switch module to AT mode
  digitalWrite(9, HIGH);
  Serial.begin(9600);
  Serial.println("Arduino is ready");
  Serial.println("Remember to select Both NL & CR in the serial
monitor");
  BTSerial.begin(38400);  // HC-05 default speed in AT command
more
}

void loop() {
// Keep reading from HC-05 and send to Arduino Serial Monitor
  if (BTSerial.available())
    Serial.write(BTSerial.read());

// Keep reading from Arduino Serial Monitor and send to HC-05
  if (Serial.available())
    BTSerial.write(Serial.read());
}
```

(2) Codes for Slave

```
//When a command is entered in the serial monitor on the computer
//the Arduino will relay it to the HC-05 and display the result.

#include <SoftwareSerial.h>
SoftwareSerial BTSerial(10, 11); // RX | TX
// Connect the HC-05 TX to Arduino pin 10
// Connect the HC-05 RX to Arduino pin 11

void setup() {
  Serial.begin(9600);
  Serial.println("Enter your commands:");

  // HC-05 default baud rate is 9600
  BTSerial.begin(9600);
}

void loop() {
// Keep reading from HC-05 and send to Arduino Serial Monitor
  if (BTSerial.available())
    Serial.write(BTSerial.read());

// Keep reading from Arduino Serial Monitor and send to HC-05
  if (Serial.available())
    BTSerial.write(Serial.read());
}


```

4. **Results**

To setup a communication between two HC-05 modules, you need to do the following things first:

- Set one HC-05 module as a master device
- Configure to pair with any address
- Cycle the power to the module

First, you need to place the HC-05 into the AT mode (hold the button switch closed while powering on). When in AT mode the LED on the HC-05 should blink on/off every second (Fig. 5.15).

- Check whether the HC-05 is in the AT mode or not. Enter "AT" and hit Send. You should receive "OK."

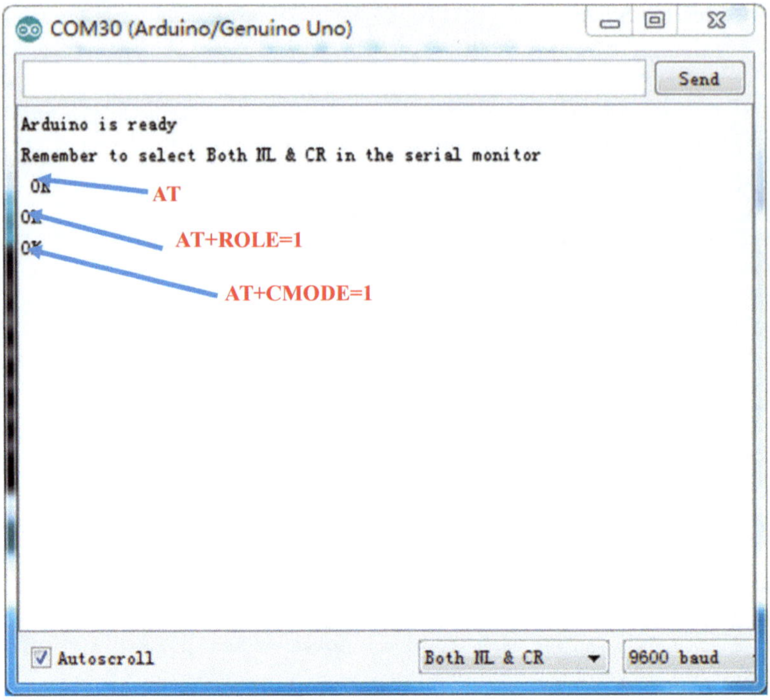

Fig. 5.15 Setting the HC-05 as a Master using the Arduino serial monitor

- Set HC-05 as a master device, Enter "AT + ROLE = 1," and hit send, you should receive another "OK."
- Configure HC-05 to pair with any address. Enter "AT + CMODE = 1" and hit send, you should receive another "OK."

Cycle the power to the master HC-05. The LED on the master HC-05 will blink twice a second as it searches and once it has connected with the slave HC-05 the LED will quickly blink twice every 2 s or so. The LED on the slave HC-05 should be constantly on.

Now, whatever you enter into one of the serial monitors will be sent by Bluetooth to the other Arduino and then displayed in the opposite serial monitor. It should be noted that you should change the baud rate from 38,400 bit to 9600 bit in the master HC-05. Or else, you will not receive the correct responses (Figs. 5.16 and 5.17).

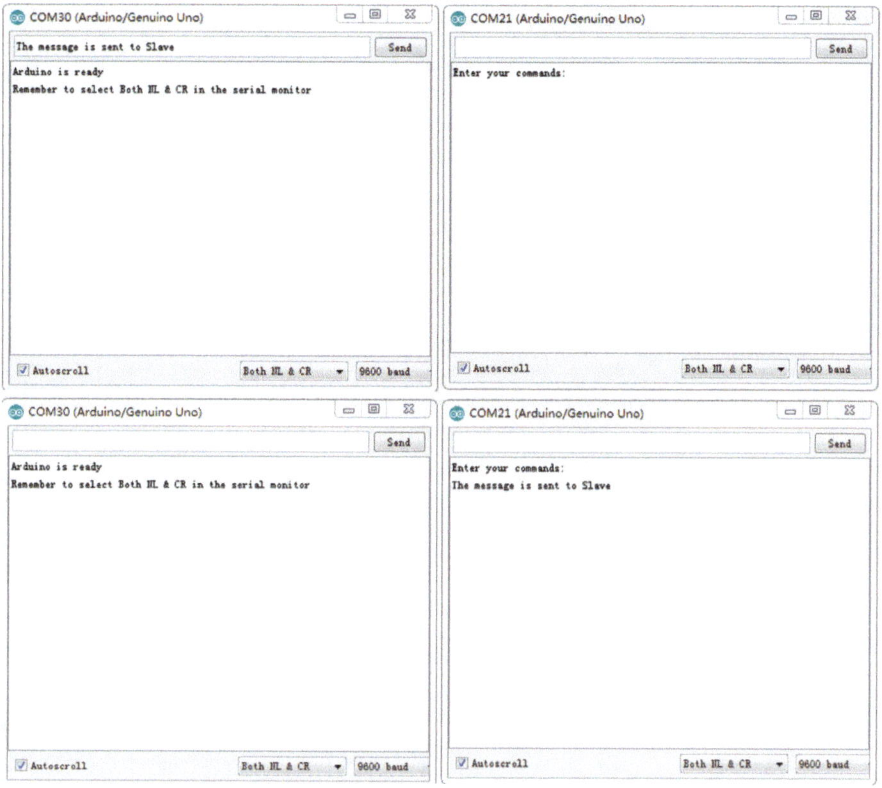

Fig. 5.16 Message from the master HC-05

5.4 GSM/GPRS Module

5.4.1 Introduction

The GSM/GPRS system is the most widely used cellular technology in use in the world today. GSM stands for Global System for Mobile Communication, which is a digital cellular technology used for transmitting mobile voice and data services. It is also sometimes referred to as 2G, as it is a second-generation cellular network. GSM supports outgoing and incoming voice calls, Simple Message System (SMS or text messaging), and data communication (via GPRS).

The GPRS (General Packet Radio Service) is a wireless packet data service that is an extension of the GSM network. It provides an efficient method to transfer data by optimizing the use of network resources. The GPRS radio resources allocator allows the provision of multiple radio channels to a single user in order to ensure a high data user rate. Furthermore, one radio channel can be shared by multiple users in order to optimize radio resources. Then, GPRS enables a high spectrum

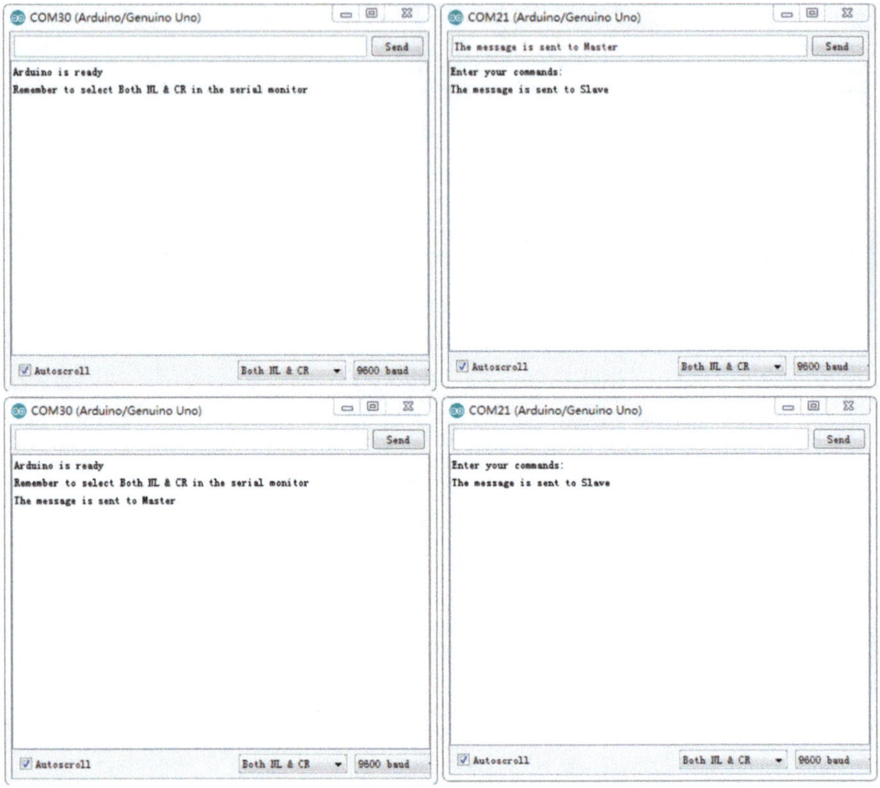

Fig. 5.17 Message from the slave HC-05

efficiency by sharing time slots between different users, supporting data rates of up to 170 kbit/s and providing a very low call setup time. Additionally, GPRS offers direct Internet Protocol (IP) connectivity in a point-to-point or a point-to-multipoint mode and provides packet radio access to external packet data networks (PDN).

There are many kinds of GSM/GPRS modules available in the market. Those GSM/GPRS modules are a ready solution for remote wireless applications, machine to machine, or user to machine and remote data communications in all vertical market applications. A GSM/GPRS module is designed for communication of a computer with the GSM and GPRS network. It requires a SIM (Subscriber Identity Module) card just like mobile phones to activate communication with the network. Furthermore, they have an IMEI (International Mobile Equipment Identity) number similar to mobile phones for their identification. The power supply circuit is also built in the module that can be activated by using a suitable adaptor. A GSM/GPRS module can perform the following operations:

(1) Receive, send, or delete SMS messages in a SIM.
(2) Read, add, and search phonebook entries of the SIM.
(3) Make, receive, or reject a voice call.
(4) Receive/send data from/to a remote location through GPRS.

The module needs AT commands, for interacting with the processor or controller, which are communicated through serial communication. These commands are sent by the controller/processor. The module sends back a result after it receives a command. Different AT commands supported by the module can be sent by the processor/controller/computer to interact with the GSM/GPRS cellular network.

5.4.2 A6 GSM/GPRS Module

In this chapter, we use the most popular A6 GSM/GPRS module as an example. The module is basically a GSM Modem connected to a PCB with different types of outputs taken from the board—TTL Output (for microcontrollers) and RS232 Output to interface directly with a PC (personal computer). The board also has pins or provisions to attach a mic and speaker, to accept +5 V or other values of power and ground connections. The features are as follows:

- Supports up to eight channel network connections.
- Low power consumption: standby is as low as 3 mA
- Wide operating temperature range.
- Dual-Band GSM/GPRS 900/1800 MHz.
- RS232/TTL interface for direct communication with computer or MCU kit.
- Configurable baud rate.
- ESD Compliance.
- Consists of a MIC and Speaker socket.
- With slide in SIM card tray.
- With Stub antenna.
- Stackable UNO headers.
- Optional power on through microcontroller.

The module shape is shown in Fig. 5.18 and has the following pins that connect to a microcontroller:

- **Antenna interface**: connected to external antenna
- **Earphones**: to answer phone calls
- **MicroSIM Card**: activate communication with the network
- **EN**: MP1584 Power chip enable pin, pull high enable power chip, pull low disenable, the foot can be used as the reset pin of the module
- **PWR**: Power on
- **REC +/REC**: Speaker positive/negative
- **MIC +/MIC**: Microphone positive/negative

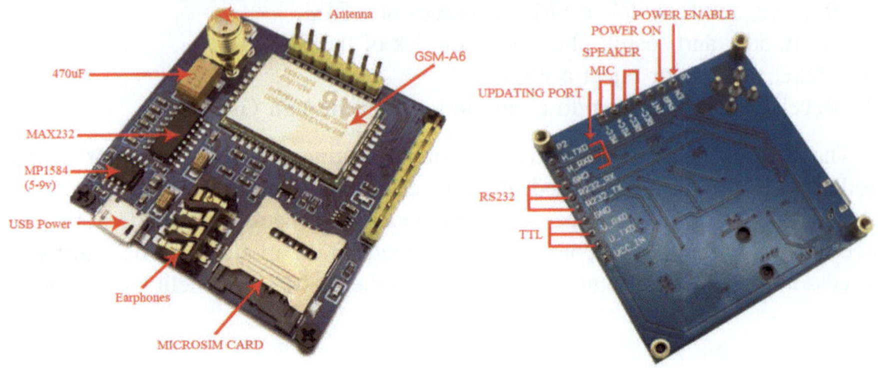

Fig. 5.18 A6 GSM/GPRS module

- **H_TXD/H_RXD**: Pinout for firmware upgrade
- **GND/R232_RX/R232_TX**: RS232 Send out/receive
- **GND/U_RXD/U_TXD**: A6 module send out/receive (TTL level)
- **VCC_IN**: Power input pin, 5–9 V

As mentioned before, the module interacts with the processor or controller by using the AT commands. Here, we just list some common AT commands (Tables 5.5 and 5.6):

5.4.3 Demonstration

In this experiment, we are going to see how to interface the **A6 GSM/GPRS Module** to **Arduino**.

1. **Components**
   - DFRobot UNO R3 board and USB cable × 1.
   - A6 GSM/GPRS Module × 1.
   - SIM card × 1.
   - GSM Antenna × 1.
   - Jumper wires × *n*.
2. **Hardware setting**
   In any case, the communication between Arduino and the GSM/GPRS module is serial. Therefore, we are supposed to use serial pins of Arduino (Rx and Tx). In this example, the pins 9 and 10 (which are PWM enabled pins) are used. This method is made possible with the SoftwareSerial Library of Arduino. SoftwareSerial is a library of Arduino that enables serial data communication through other digital pins of Arduino. The library replicates hardware functions and handles the task of serial communication. The circuits is shown in Fig. 5.19:

Table 5.5 AT commands for SMS

Command	Description	Response	Result
AT + CMGF	Set the module message mode, set either at PDU(0) or text mode (1)	OK	–
AT + CNMI	Set the new message remind, for example AT + CNMI = 2,1,	+CMTI: “SM”,2	When set is on and message box is NOT full, message is stored at position 2
AT + CMGS	Send message, send 180 bytes at GSM mode, or 70 Chinese characters at UCS2 mode, AT + CMGS = “15805298357”	Will return “>” and then type message, then end up with HEX value 1A (ox1A, “CTRL + Z”), send 1B to cancel “ESC”	And finally return: +CMGS:156, in which 156 has meaning
AT + CMGR	Read message, for example, AT + CMGR = 1 to read message at position 1	–	–
AT + CMGL	List all messages received on the GSM modem, for example, AT + CMGL = “ALL”	+CMGL: <index>, <stat>, <oa>, [<alpha>], [<scts>] <CR> <LF> <data> <CR> <LF> ... OK	+CMGL: 1, “REC UNREAD”, “+ 31628870634”, “11/01/09,10:26:26 + 04” This is text message 1 +CMGL: 2,”REC UNREAD”,” + 31628870634”, ”11/01/09,10:26:49 + 04” This is text message 2 OK
AT + CMGD	Delete message, for example AT + CMGD = 1 to delete message at position 1	OK	–

Table 5.6 AT commands for GPRS

Command	Description	Response	Result
AT + CGATT	Attach (1) or disattach (0) GPRS service, for example, AT + CGATT = 1	OK	–
AT + CGACT = 1,1	Activate (AT + CGACT = 1,1) and deactivate (AT + CGACT = 0,1) PDP context		
AT + CIPSTART	Initiate a connect or setup a UDP port, for example, AT + CIPSTART = "TCP", "180.120.52.222", "8086", to connect to 180.120.52.222 at port 8086	Connect OK	–
AT + CIPSEND	Send data, module will return ">", send max 1352 bytes and end up with 1A same as SMS	Send OK	–
AT + CIPCLOSE	Close current TCP/UDP connection status	–	–

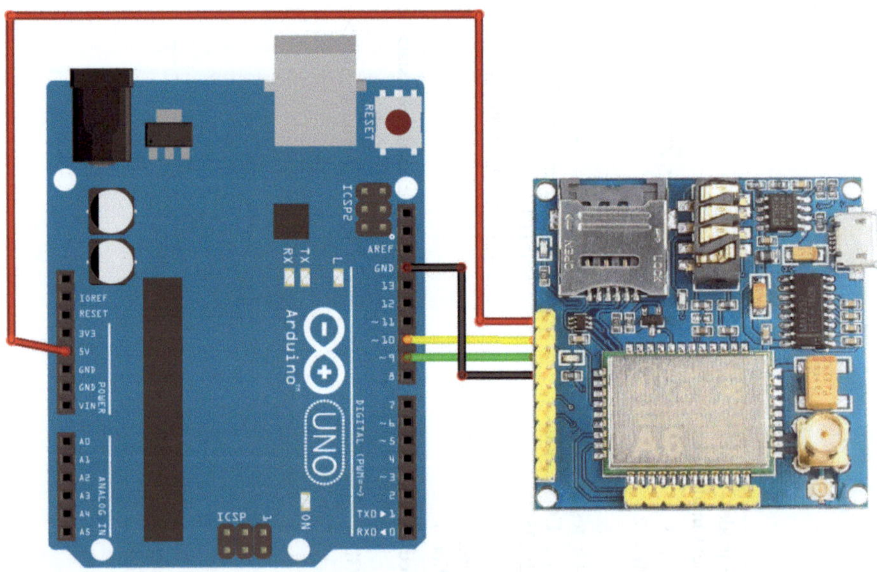

Fig. 5.19 The diagram of layout of A6 GSM/GPRS module

3. **Sample Codes**

(1) GSM Example

```
#include <SoftwareSerial.h>

SoftwareSerial mySerial(9, 10);

void setup() {
  mySerial.begin(9600);   // Setting the baud rate of GSM Module
  Serial.begin(9600);    // Setting the baud rate of Serial
Monitor (Arduino)
  delay(100);
}

void loop() {
  if (Serial.available() > 0)
    switch (Serial.read())
    {
      case 's':
        SendMessage();
        break;
      case 'r':
        RecieveMessage();
        break;
    }
  if (mySerial.available() > 0)
    Serial.write(mySerial.read());
}

void SendMessage() {
  mySerial.println("AT+CMGF=1");  //Sets the GSM Module in Text
Mode
  delay(1000);                    // Delay of 1000 milli seconds
or 1 second
  mySerial.println("AT+CMGS=\"+9115805298357\"\r"); // mobile
number
  delay(1000);
  mySerial.println("I am SMS from GSM Module"); // The SMS text
you want to send
  delay(100);
  mySerial.println((char)26);// ASCII code of CTRL+Z
  delay(1000);
}

void RecieveMessage() {
  mySerial.println("AT+CNMI=2,2,0,0,0"); // AT Command to receive
a live SMS
  delay(1000);
}
```

The steps of booting the A6 GSM/GPRS module are as follows:

(1) Insert the SIM card into the GSM/GPRS module and lock it.
(2) Connect the adapter to the GSM/GPRS module and turn it ON.
(3) Wait for some time (say 1 min) and see the blinking rate of "status LED" or "network LED" (GSM/GPRS module will take some time to establish a connection with the mobile network).
(4) Once the connection is established successfully, the status/network LED will blink continuously every 3 s. You may attempt to make a call to the mobile number of the SIM card inside the GSM/GPRS module. If you hear a ring back, the GSM/GPRS module has successfully established a network connection.

In this example, the communication is triggered by the user input. The program seeks user input via the **serial monitor** of Arduino. If the input is "**s**," the program will invoke a function to **send an SMS from the GSM module**. If the user input is "**r**," the program will invoke a function to **receive a live SMS** from the GSM module and display it on the serial monitor of Arduino (Fig. 5.20).

(2) GPRS Example

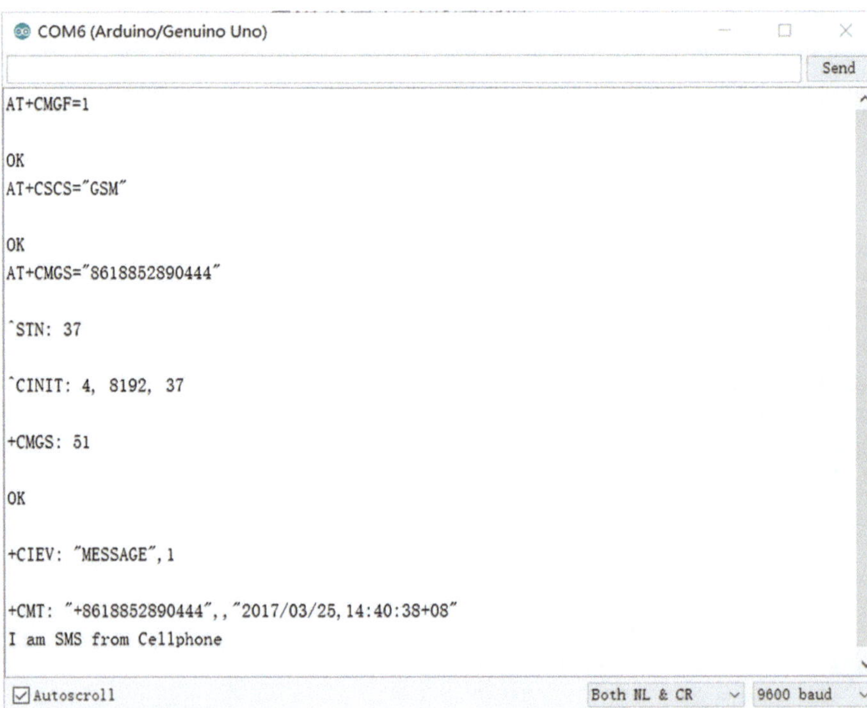

Fig. 5.20 A6 GSM/GPRS result in Arduino serial monitor

5.5 Wi-Fi Module

5.5.1 Introduction

Wi-Fi, or Wireless Fidelity, is a term that is used generically to refer to any product or service using any type of 802.11 technology. Wi-Fi networks operate in the unlicensed 2.4–2.5 GHz radio bands, with an 11 Mbps (802.11b) or 54 Mbps (802.11a) data rate, respectively. Wi-Fi technology may be used in a variety of scientific, industrial, commercial, and consumer applications. Many devices can use Wi-Fi, e.g., personal computers, video-game consoles, smartphones, digital cameras, tablet computers, and digital audio players. These devices can all connect to a network resource via a wireless network access point. The following list summarizes some of the benefits of a Wi-Fi network.

- **Wireless Ethernet**. Wi-Fi is an Ethernet replacement. Wi-Fi and Ethernet, both IEEE 802 networks, share some core elements.
- **Extended Access**. The absence of wires and cables extends access to places where wires and cables cannot go or where it is too expensive for them to go.
- **Cost Reduction**. As mentioned above, the absence of wires and cables brings down cost. This is accomplished by a combination of factors, the relatively low cost of wireless routers, and cost-savings from not trenching, drilling, and other methods that may be necessary to make physical connections.
- **Mobility**. Wires tie you down to one location. Going wireless means you have the freedom to change your location without losing your connection.
- **Flexibility**. Extended access, cost reductions, and mobility create opportunities for new applications as well as the possibility of creating new solutions for legacy applications.

5.5.2 Wi-Fi Module

There are many different kinds of Wi-Fi Modules & Solutions available when you do your search. The vendors or module makers usually categorize the modules by many parameters including data rate, range, RF band, certification and packaging type, etc. In this section, a DFRobot Wi-Fi module v2.2 (TEL0047) is chosen as an example (shown in Fig. 5.21, and its functions are shown in Table 5.7).

The module uses WizFi210, which is a low power-consuming Wi-Fi module that is applied with dynamic power management technology. When the Wi-Fi module needs to be on but not operate, it can be set to Standby mode, a low power mode, and the module simply needs to be woken up when it has to work. It provides bridging from TTL serial port communication to IEEE802.11b/g/n wireless communication. Therefore, any device with TTL serial ports can be easily connected

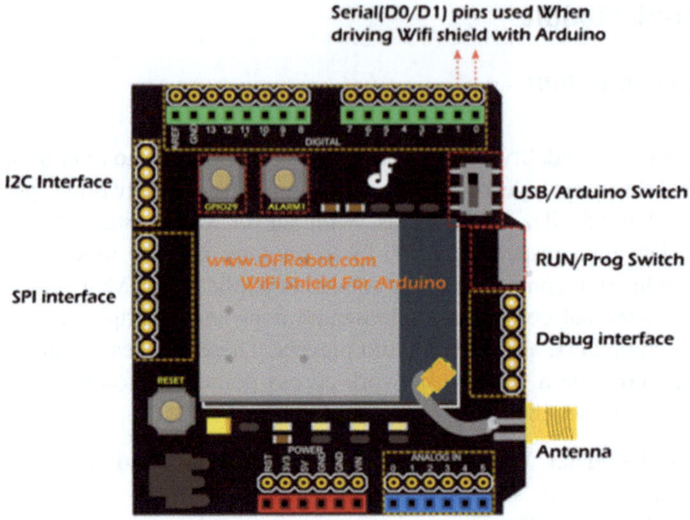

Fig. 5.21 Wi-Fi module

Table 5.7 The instruction for the function of Wi-Fi module v2.2

Interface	Functions
USB/Arduino Switch	USB: Config Wi-Fi module via PC comm port directly Arduino: Enable the communication between Arduino and Wi-Fi module
RUN/Prog Switch	RUN: Enable the USB or Arduino to drive the Wi-Fi module Prog: Disable the Wi-Fi module for programming Arduino via USB
Debug interface	Drive and config the Wi-Fi module directly by using the FTDI or USB serial light adapter

with this Wi-Fi module and controlled and managed remotely through a wireless network. Different kinds of communication protocols and encryption algorithms are integrated with the module. The Arduino architecture enables you to easily integrate this module into any Arduino based project and achieve Wi-Fi communication via UART by sending AT commands. The specifications are as follows:

- Radio Protocol: IEEE 802.11b/g/n compatible
- Supported Data Rates: 11, 5.5, 2, 1 Mbps (IEEE 802.11b)
- Modulation: DSSS and CCK
- RF Operating Frequency: 2.4–2.497 GHz
- Antenna Options: Chip antenna and U.FL connector for external antenna
- Networking Protocols: UDP, TCP/IP (IPv4), DHCP, ARP, DNS, HTTP/HTTPS Client and Server (*)
- Power Consumption (Typical): Standby = 34.0 μA Receive = 125.0 mA Transmit = 135.0 mA
- RF Output Power (Typical): 8 ± 1 dBm

- Security Protocols: WEP, WPA/WPA2-PSK, Enterprise, EAP-FAST, EAP-TLS, EAP-TTLS, PEAP
- I/O Interface: UART, SPI(*), I2C(*), WAKE, ALARM, GPIOs
- Working Voltage: 5 V
- Chip working voltage: 3.3 V
- Dimensions (except Antenna): 59 × 54 mm

The following AT commands are a quick review of the essentials needed to get your Wi-Fi module connected to your network with a UDP server running on port 4000. This is not meant to be an all-encompassing tutorial. If you would like to make a more complex setup, please refer to the WizFi datasheet provided in the documents section of this wiki for a complete list of AT commands and their description (Table 5.8).

Table 5.8 The AT commands for configuration

Command	Response	Description
AT	OK	Enter at mode
AT + WD	AT + WD OK	Disassociate from previous connection
AT + WWPA = 1234	AT + WWPA = 1234 OK	Set wireless password
AT + NDHCP = 1	AT + NDHCP = 1 OK	Enable DHCP settings. Auto assign IP address. You might want to disable this option "0" and set your own IP address
AT + WA = YOURSSID		Define router's "SSID"
AT + NSTAT = ?		Current wireless and network config. **Note: write down the reported IP address. You will need it later**
AT + WSTATUS		Adapter reports the current network config to serial host
AT + DNSLOOKUP = baidu.com		Test your connection to the internet. If successful it will return baidu's IP address
AT + NSTCP = 5000	<Port>	Set TCP server at port 5000
AT + NSUDP = 4000	<Port>	Set UDP server at port 4000
AT + CID = ?	<CID>	Returns the current CID configuration
ATC1	ATC1 OK	Set to auto connect at restart
AT&W0	AT&W0 OK	Save settings to profile "0"
ATA		Connect

5.5.3 Demonstration

In this experiment, we are going to see how to interface the **Wi-Fi Module** to **Arduino**.

1. **Components**
   - DFRobot UNO R3 board and USB cable × 1.
   - DFRobot Wi-Fi module v2.2 (TEL0047) × 1.
   - Router × 1.
   - Antenna × 1.
   - Jumper wires × 1.
2. **Hardware setting**
   The Wi-Fi module v2.2 is an expansion board of Arduino, which can be plugged into the UNO directly (Fig. 5.22).
   Before implementation, you should configure the Wi-Fi module first,

   (1) Program a simple led blinking sample code for your Arduino UNO R3 first to avoid a serial communication conflict between the USB port and the Wi-Fi module.
   (2) Stack the Wi-Fi module to the Arduino UNO R3 and connect the antenna to the Wi-Fi module.
   (3) Set the "USB/Arduino" Switch to the **USB** side to enable the PC com port to send AT commands to the Wi-Fi module directly.
   (4) Set the "Run/Prog" Switch to the **RUN** side.
   (5) Plugin the A to B USB to the Arduino microcontroller. The "POWER" LED turns on at the same time.

Fig. 5.22 A diagram of the layout of the Wi-Fi module v2.2 and UNO R3

Then, the Serial monitor included in Arduino IDE is used to configure the Wi-Fi module. Please set the baud rate to 115,200 bps for both NL and CR.

(1) Open the com port of your Arduino UNO. Enter "AT" to test the communication between the Wi-Fi module and the USB port.
(2) Enter the "AT + WS" command to scan the network and obtain a list of found networks (shown in Fig. 5.23).
(3) Setting the Wi-Fi module according to the router information. List the sample command below (Fig. 5.24).
(4) Then, save the setting (at&w0). Otherwise, all the settings above will be cleared after resetting.
(5) After getting the feedback correctly from the Wi-Fi module, we finish the setting for the Wi-Fi shield and create a TCP server in the local network. Next, the indicator LEDs (marked "STW" and "ASSOC") turn on to indicate the connection with the router.

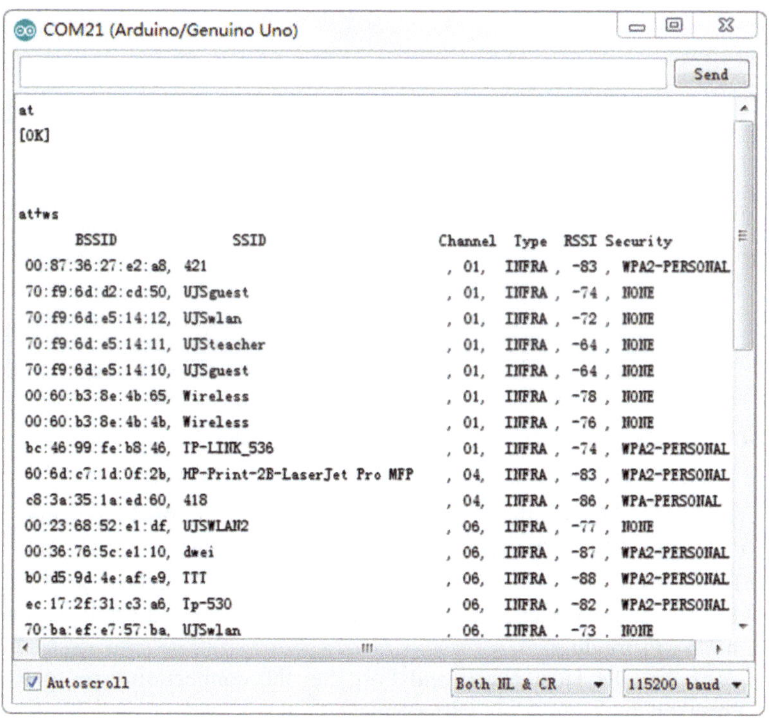

Fig. 5.23 Scanning the network using the Arduino serial monitor

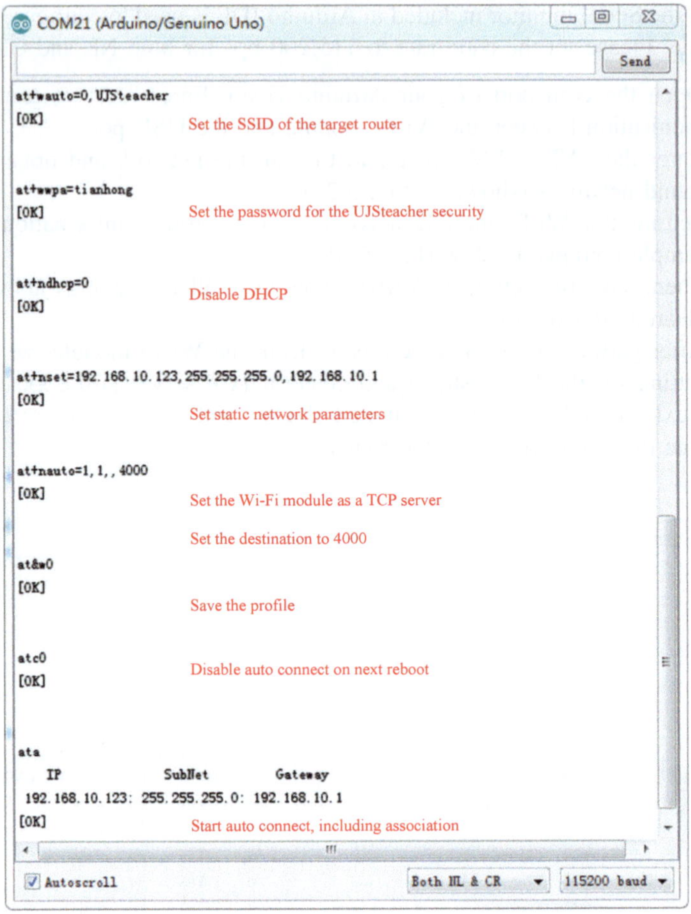

Fig. 5.24 Setting Wi-Fi module using the Arduino serial monitor

(6) Next, we create a client and send a message to our Wi-Fi module.
Connect to the TCP server from the Wi-Fi module.
The PuTTy (http://www.chiark.greenend.org.uk/~sgtatham/putty/download.html) is chosen to work as a TCP client and send commands to the Wi-Fi module.
Configure the Host Name and Port. Set the connection type to Raw and Press open. Then, you will connect to the server created by the Wi-Fi module (Fig. 5.25).
Now, you could send commands via the TCP client simulated by putty to the Wi-Fi module. Further, putty will receive the strings sent from the Serial monitor as well.

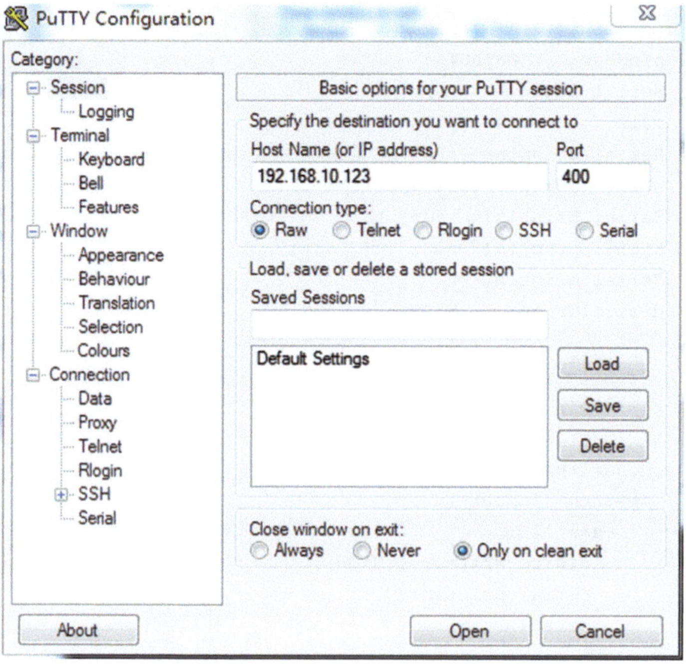

Fig. 5.25 Configure PuTTy software

1. **Sample codes**

Once you have verified that your Shield is connected to the internet you can upload this sketch to your Arduino by simply removing the 2 jumpers, and replacing them once the sketch has been uploaded. There is no need to change the settings for this, so just replace them on the "USB" mode.

You can test the Wi-Fi communication with this shield by removing the USB cable and providing external power. It is recommended that you provide 7.5 Vdc and 1 A to ensure that the Wi-Fi Radio has enough power.

You should now be able to connect to your shield's IP address and the specified UDP port using PuTTy.

Pressing 1 and 0 should turn LED 13 on and off, while reporting back to you on the PuTTy screen.

```
void setup() {
  pinMode(13, OUTPUT);
  Serial.begin(115200);
  delay(100);
  Serial.println("Press any key to continue");
  while (Serial.available() == 0);
  Serial.println("  Lets test the DFRobot WiFi module");
  Serial.println("Press 1 to light the LED, and 0 to turn it off");
  Serial.println("Entery: ");
  digitalWrite(13, HIGH);
}

void loop() {
  if (Serial.available()) {
    char input = Serial.read();
    switch (input) {
      case '1':
        digitalWrite(13, HIGH);
        Serial.println("ON");
        delay(500);
        break;
      case '0':
        digitalWrite(13, LOW);
        Serial.println("OFF");
        delay(500);
        break;
    }
  }
}
```

Part II
Case Studies

Chapter 6
PM2.5/Air Quality Monitor Using Arduino

6.1 Introduction

Dust, known as "airborne particles", refers to solid particles that are suspended in air (diameter $\leq$ 100 μm). Among them, dust with diameter < 2.5 μm, known as "fine particulate matter", is referred to as PM2.5. After being inhaled by the human body, PM2.5 can directly enter the blood through the bronchi and alveoli, which are phagocytized by macrophage. Their long-term stay in the alveoli exerts a negative impact on the human cardiovascular, nervous system, and other organs, posing a threat to people's health. Consequently, indoor air quality has become a great concern.

In this chapter, an indoor air quality monitor based on Arduino is developed, which integrates a thermometer, a humidity meter, an air quality monitor, and a HEPA filter into one single system, in order to help users gauge and improve indoor air quality. Furthermore, Bluetooth communication is also used in this system. The Bluetooth communication system enables a user to remotely control and observe the air quality monitor.

6.2 System Design

The indoor air quality monitor mainly consists of a dust sensor module, humidity, and temperature sensor module, Arduino UNO R3, a liquid-crystal display module, electronic speed control (ESC) module, suction fan and air HEPA filters, a wireless transmitting and receiving module (Bluetooth), and power supply circuit, as shown in Fig. 6.1.

First, the dust particle information is converted into an electrical signal by the dust sensor, and after passing the preamplifier circuit of the sensor, it is converted into digital electrical signals by the A/D conversion circuit, which is fed to Arduino

T. Pan and Y. Zhu, *Designing Embedded Systems with Arduino*,
DOI 10.1007/978-981-10-4418-2_6

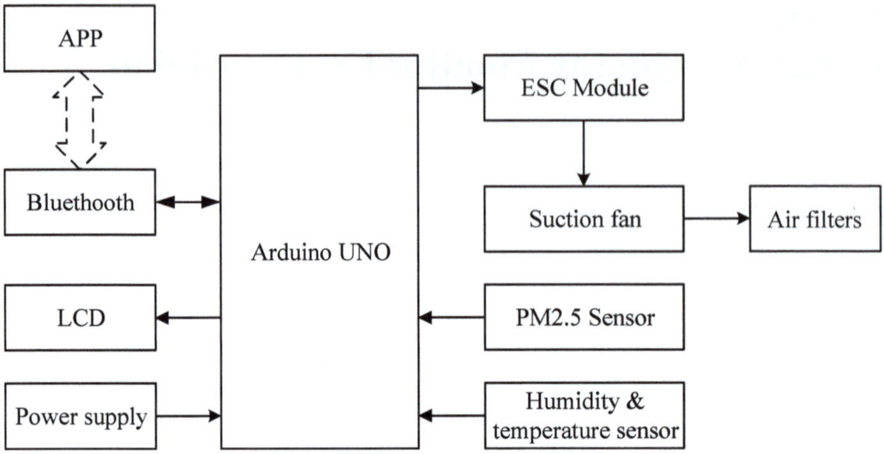

Fig. 6.1 Structure diagram of the indoor air quality monitor

for processing and analysis. Arduino displays the processed and analyzed data through the LCD. When Arduino determines that the dust concentration exceeds preset indicators, the suction fan operates using the ESC circuit, and the indoor air is filtered through the HEPA filter. The Bluetooth module is used to realize sharing and real-time transmission of data and information.

6.2.1 Air Quality Sensor (SEN0177)

An air quality sensor can detect the particle concentration in an indoor environment and communicate the same to Arduino. Therefore, Arduino can obtain air quality information. In this project, the SEN0177 dust sensor is used, which can obtain the number of suspended particulate matter in a unit volume of air within 0.3–10 μm, namely the concentration of particulate matter, and output it to a digital interface; it can also output the quality data per particle (shown in Fig. 6.2).

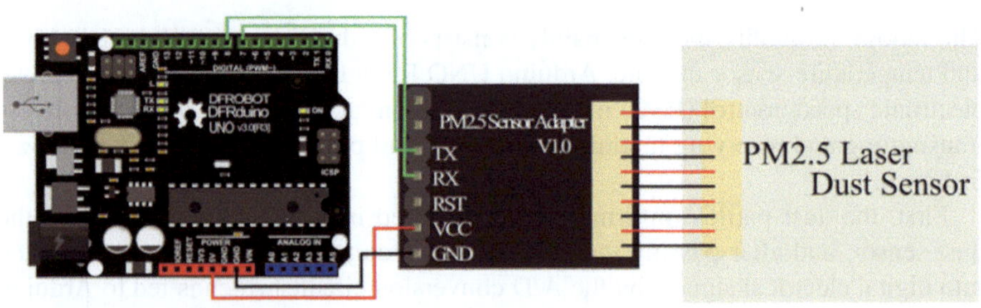

Fig. 6.2 Connecting diagram between UNO and SEN0177

It can be seen that the PM2.5 laser dust sensor is connected to the Arduino UNO through the PM2.5 sensor adapter with TXD and RXD pins. The RS232 on-board the UNO (RXD, TXD) is used to connect with the Bluetooth module. Therefore, the software serial port is used to connect with the PM2.5 sensor adapter. Before writing codes, the SoftwareSerial library (http://arduiniana.org/NewSoftSerial/NewSoftSerial12.zip) should be included first. The SoftwareSerial library has been developed to allow serial communication on other digital pins of the Arduino, using software to replicate the functionality. It is possible to have multiple software serial ports with speeds of up to 115,200 bps. A parameter enables inverted signaling for devices that require that protocol.

The packet length of the PM2.5 sensor adapter is fixed at 32 bytes and the information is shown in Table 6.1.

Table 6.1 The packet information of SEN0177

Bits	Information
Start character 1	0 × 42 (fixed bit)
Start character 2	0 × 4d (fixed bit)
Frame length 16-byte	Frame length = 2 * 9 + 2 (data + check bit)
Data 1, 16-byte	Concentration of PM1.0, μg/m^3
Data 2, 16-byte	Concentration of PM2.5, μg/m^3
Data 3, 16-byte	Concentration of PM10.0, μg/m^3
Data 4, 16-byte	Internal test data
Data 5, 16-byte	Internal test data
Data 6, 16-byte	Internal test data
Data 7, 16-byte	The number of particulate of diameter above 0.3 μm in 0.1 L of air
Data 8, 16-byte	The number of particulate of diameter above 0.5 μm in 0.1 L of air
Data 9, 16-byte	The number of particulate of diameter above 1.0 μm in 0.1 L of air
Data 10, 16-byte	The number of particulate of diameter above 2.5 μm in 0.1 L of air
Data 11, 16-byte	The number of particulate of diameter above 5.0 μm in 0.1 L of air
Data 12, 16-byte	The number of particulate of diameter above 10.0 μm in 0.1 L of air
Data 13, 16-byte	Internal test data
Check Bit for data sum, 16-byte	Check Bit = Start character 1 + Start character 2 + …all data

The reading codes are shown as follows

```
#include <Arduino.h>
#include <SoftwareSerial.h>
SoftwareSerial mySerial(7, 8); //create softwareSerial
#define rx_Pin 7 //define the software RXD pin
#define tx_Pin 8 //define the software TXD pin
#define PMArrayLenth 31 //0x42 +31 bytes = 32 bytes
unsigned char buf[PMArrayLenth];

int PM01Value = 0; //define PM1.0 value
int PM2_5Value = 0; //define PM2.5 value
int PM10Value = 0; //define PM10 value

void setup() {
  mySerial.begin(9600); //configure the baudrate
  mySerial.setTimeout(1500); //set the Timeout to 1500ms, longer
than the data transmission periodic time of the sensor
}

void loop() {
  //start to read when detect 0x42
  if (mySerial.find(0x42)) {
    mySerial.readBytes(buf, PMArrayLenth);
    if (buf[0] == 0x4d) {
      if (checkValue(buf, PMArrayLenth)) {
        PM01Value = transmitPM01(buf);
        PM2_5Value = transmitPM2_5(buf);
        PM10Value = transmitPM10(buf);
      }
    }
  }
}

char checkValue(unsigned char *thebuf, char leng) {
  char receiveflag = 0;
  int receiveSum = 0;
  for (int i = 0; i < (leng - 2); i++) {
    receiveSum = receiveSum + thebuf[i];
  }
  receiveSum = receiveSum + 0x42;
  if (receiveSum == ((thebuf[leng - 2] << 8) + thebuf[leng - 1]))
{ //check the serial data
    receiveSum = 0;
    receiveflag = 1;
  }
  return receiveflag;
}

//count PM1.0 value of the air detector module
int transmitPM01(unsigned char *thebuf) {
  int PM01Val;
  PM01Val = ((thebuf[3] << 8) + thebuf[4]);
  return PM01Val;
}

//count PM2.5 value of the air detector module
```

```
int transmitPM2_5(unsigned char *thebuf) {
  int PM2_5Val;
  PM2_5Val = ((thebuf[5] << 8) + thebuf[6]);
  return PM2_5Val;
}

//count PM10 value of the air detector module
int transmitPM10(unsigned char *thebuf) {
  int PM10Val;
  PM10Val = ((thebuf[7] << 8) + thebuf[8]);
  return PM10Val;
}
```

6.2.2 *Temperature and Humidity Sensor (DHT11)*

The temperature sensor can transform the temperature information into voltages. After analog to digital conversion, the temperature information can be obtained. The humidity sensor can transform the humidity information into electrical signal. After analog to digital conversion, the humidity information can be obtained. Here, the DHT11 module (DFR0067) is used. Please refer to Sect. 3.4 in detail.

6.2.3 *Liquid-Crystal Display*

The liquid-crystal display is an optional approach to display the data of the temperature, humidity, and air quality information. Here, LCD1602 is used, which is a 2×16 LCD. The connecting diagram is shown in Fig. 6.3.

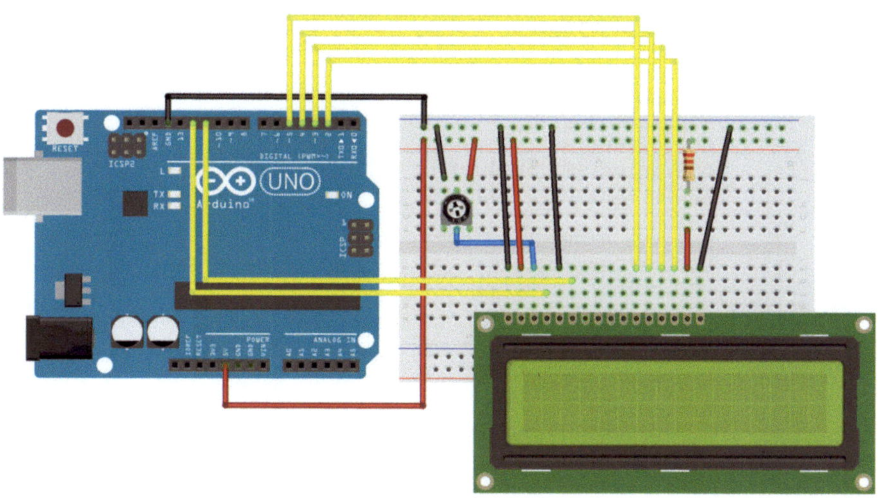

Fig. 6.3 Connecting diagram between UNO and LCD1602

To wire your LCD screen to your board, connect the following pins:

- LCD RS pin to digital pin 12
- LCD Enable pin to digital pin 11
- LCD D4 pin to digital pin 5
- LCD D5 pin to digital pin 4
- LCD D6 pin to digital pin 3
- LCD D7 pin to digital pin 2

Additionally, wire a 10k pot to +5 V and GND, with its wiper (output) to LCD screens VO pin (pin3). A 220 Ω resistor is used to power the backlight of the display, usually on pin 15 and 16 of the LCD connector.

The codes are as follows. Here, we only provide the display codes, you need to obtain the other detection codes (dust, humidity, and temperature detection) from previous sections.

```
#include <dht11.h>
#include <LiquidCrystal.h>
#include <dht11.h>
#include <Arduino.h>
#define DHT11_PIN 6
dht11 DHT;
LiquidCrystal LCD(12, 11, 5, 4, 3, 2);
#define PMArrayLenth 31  //0x42 + 31 bytes = 32 bytes
unsigned char buf[PMArrayLenth];
int PM01Value = 0; //define PM1.0 value
int PM2_5Value = 0; //define PM2.5 value
int PM10Value = 0; //define PM10 value

void setup()
{
  Serial.begin(9600);
  LCD.begin(16, 2);
  Serial.setTimeout(1500);
}

void loop() {
  LCD.setCursor(0, 0); //display PM2.5
  LCD.print("PM2.5=");
  LCD.print(PM2_5Value);
  LCD.print("ug/m3");

  int chk = DHT.read(DHT11_PIN);
  LCD.setCursor(0, 1); // display temperature
  LCD.print("TMPR=");
  LCD.print(DHT.temperature);
  LCD.print("C");

  LCD.setCursor(9, 1);
  LCD.print("HUM=");
  LCD.print(DHT.humidity); // display humidity
  LCD.print("%");
}
```

The result is shown as follows (Fig. 6.4).

Fig. 6.4 Detecting results

6.2.4 *Servo*

According to air quality information obtained by the dust sensor, Arduino can control the suction fan to rotate at a certain speed. Then, the indoor air is filtered through the HEPA filter (shown in Fig. 6.5).

The suction fan is connected with ESC which are PWM controllers. The ESC generally accepts a nominal 50 Hz PWM servo input signal whose pulse width varies from 1 to 2 ms. When supplied with a 1 ms width pulse at 50 Hz, the ESC, responds by turning off the motor attached to its output. A 1.5 ms pulse-width input signal results in a 50% duty cycle output signal that drives the motor at approximately half speed. When presented with a 2.0 ms input signal, the motor runs at full speed owing to the 100% duty cycle (on constantly) output (Fig. 6.6).

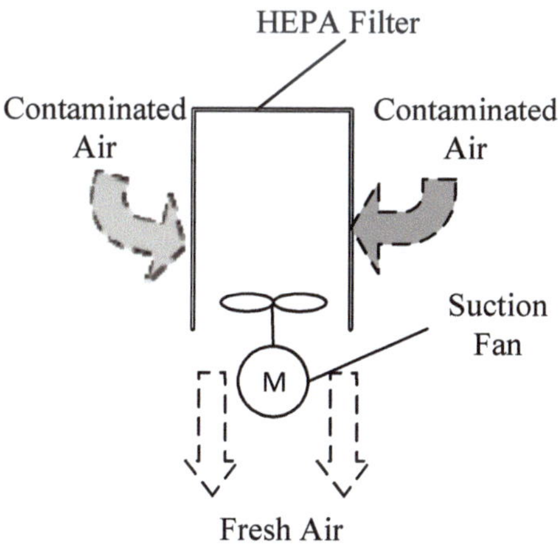

Fig. 6.5 The filtering principle of air indoor

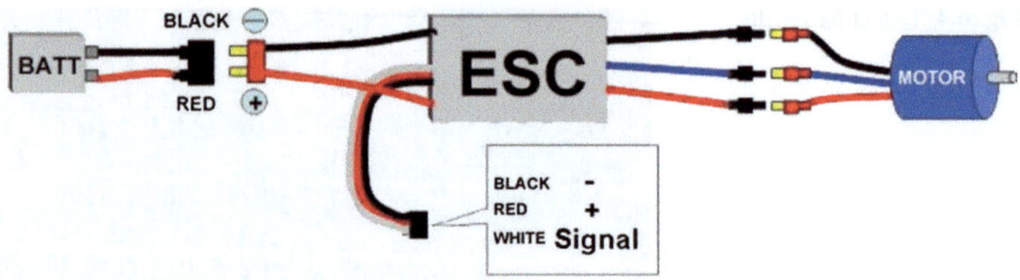

Fig. 6.6 Connecting diagram of ESC

The codes are shown as follows:

```
#include <Servo.h>
Servo myservo;
int ESC_Pin = 9; // define the ESC output pin
int pulsewidth; // define the pulse width
int temp = 0; //test motor

// define pulse function
void servopulse(int ESC_Pin, int val) {
  int myangle;
  myangle = map(val, 0, 180, 500, 2480); // all from 0-180 to
500-2480
  digitalWrite(ESC_Pin, HIGH); // set HIGH
  delayMicroseconds(myangle); // delay
  digitalWrite(ESC_Pin, LOW); // set LOW
  delay(20 - val / 1000); // Delay
}
void setup() {
  pinMode(ESC_Pin, OUTPUT);
  for (int i = 0; i <= 110; i++) {
    servopulse(ESC_Pin, 150); //set max speed about 2s
  }
  for (int i = 0; i <= 55; i++) {
    servopulse(ESC_Pin, 20); //set min speed about 1s
  }
  for (int i = 0; i <= 150; i++) {
    servopulse(ESC_Pin, 31); //control the motor
  }
  myservo.attach(ESC_Pin); //define motor i/o
}
void loop() {
  double len = 700 + temp ; // len =0.7*1000~0.95*1000
  myservo.writeMicroseconds(len); // control motor
}
```

The key point in this program is to **set the throttle range**. You would likely have to do this once. Once the throttles are set, the values are persistent. The steps are as follows:

1. Set the initial throttle position to high.
2. Turn on the ESC. It is very important that ESC receives a high signal on the PWM pin when it is turned on. You should send a high signal and wait for a user input, so that you have time to turn the ESC on. When turned on, the ESC will emit a 1-2-3 (three tones of different frequencies in rapid succession).
3. The ESC will emit a beep-beep tone, indicating that the high throttle range has been set.
4. As soon as you hear the beep-beep, you need to send a low throttle signal. The ESC will respond by emitting several beeps (the number depends on the input voltage of the power supply connected to the ESC). These beeps will be followed by a long beep, indicating that the low point has been correctly set.

Furthermore, you should include the servo library in your code before coding; the download link is http://playground.arduino.cc/uploads/ComponentLib/SoftwareServo.zip.

6.2.5 Bluetooth (HC-05)

In this design, the HC-05 module is used to setup the communication between mobile phones and the monitor through Bluetooth. Bluetooth is a wireless technology standard for exchanging data over short distances (using short-wavelength UHF radio waves in the ISM band from 2.4 to 2.485 GHz) from fixed and mobile devices, and building personal area networks (PANs). The range is approximately 10 m (30 ft).

Using HC-05, the temperature, humidity, and air quality information are sent to an App so that users can read these information on their mobile phones. The HC-05 connection diagram is shown as follows (Fig. 6.7):

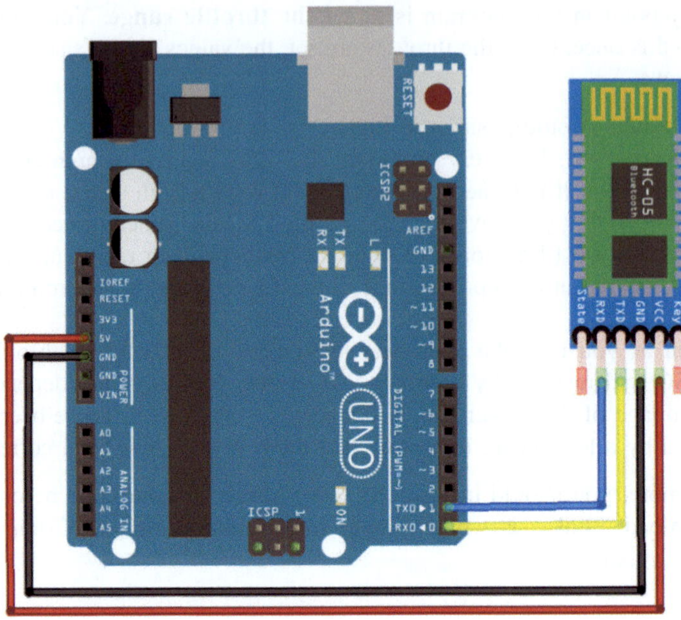

Fig. 6.7 Bluetooth connecting diagram

```
#include "Arduino.h"
#include <dht11.h> //temperature and humidity
dht11 DHT;
int PM01Value = 0; //define PM1.0 value
int PM2_5Value = 0; //define PM2.5 value
int PM10Value = 0; //define PM10 value
int HumValue = 0; //define humidity value
int TmprValue = 0; //define temperature value
unsigned int printInterval = 2000; // communication interval is 2s
unsigned long printTime;

void setup () {
  Serial.begin(9600);
}
void loop () {
  if (millis() - printTime >= printInterval) {
    printTime = millis();
    Serial.println(PM01Value);
    Serial.println(PM2_5Value);
    Serial.println(PM10Value);
    Serial.println(TmprValue);
    Serial.println(HumValue);
  }
}
```

6.2.6 Software Development

Figure 6.8 shows the flow chart of the monitor. Once the sensor readings are acquired, they are encoded into the data message for data transmission and display. Data from the temperature and humidity sensor, dust sensor are sampled every 2 s owing to the response time. Then, data messages are shown on the LCD display and transmitted to a mobile phone through Bluetooth. Based on the detected PM2.5 value, the suction fan will rotate at variable frequencies. Of course, the suction fan will stop when the PM2.5 is less than 70.

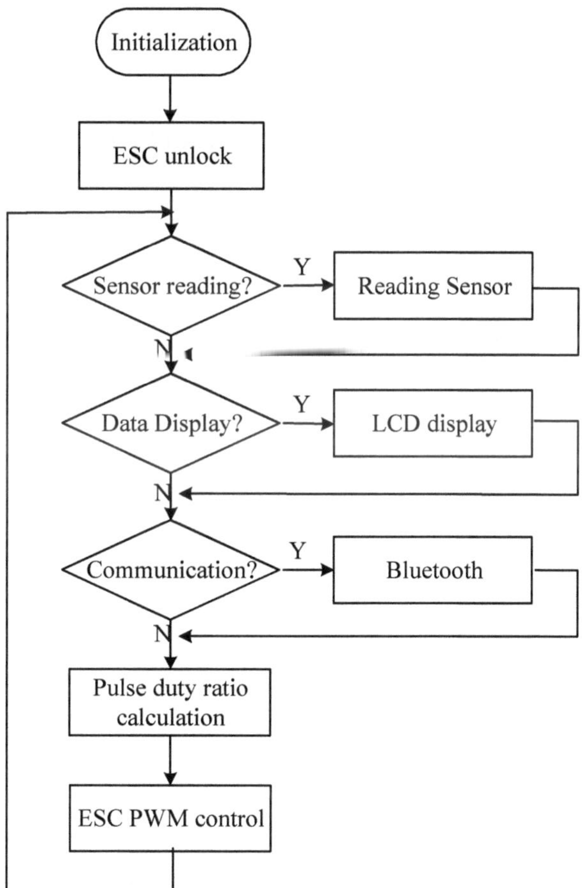

Fig. 6.8 Flowchart of indoor air quality monitor

6.3 Production Demonstration

6.3.1 Components

- DFRobot UNO R3 board and USB cable × 1.
- DFR0067 (DHT11 Temperature and Humidity Sensor V2) × 1.
- HC-05 (Bluetooth) × 1.
- SEN0177 (Air Quality Sensor) and its adapter × 1.
- Skywalker-40A (ESC) × 1.
- Brushless motor × 1.
- LCD1602 × 1.
- HEPA filter × 1.
- Jumper wires × *n*.

6.3.2 UNO R3 Digital Pinouts Are as Follows

HC-05 module
0—Bluetooth TXD pin
1—Bluetooth RXD pin
LCD 1602 module
2—LCD D7 pin
3—LCD D6 pin
4—LCD D5 pin
5—LCD D4 pin
11—LCD Enable pin
12—LCD RS pin
DHT11 module
6—Temperature and Humidity input pin
Skywalker-40A module
9—ESC output pin
SEN0177 and its adapter
7—Air quality sensor adapter TXD pin
8—Air quality sensor adapter RXD pin

6.3.3 Results

The product is shown in Fig. 6.9

The Arduino indoor environment monitor can also send information to a mobile phone using Bluetooth. The App is designed by using E4A tools (http://bbs.e4asoft.com/thread-33742-1-1.html). Pair the Bluetooth device with your mobile phone and then run the application. All environmental data are encoded into data messages in an agreed format and transmitted to the App. Figure 6.10 shows the application front view.

Fig. 6.9 The photo of air quality monitor

Fig. 6.10 View of the app

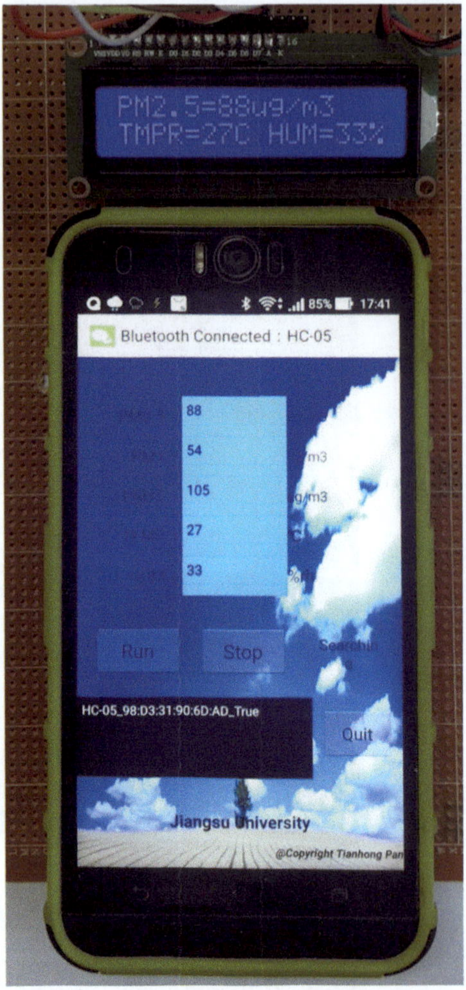

6.3.4 Codes

```
#include <LiquidCrystal.h>
LiquidCrystal lcd(12, 11, 5, 4, 3, 2); //LCD pin
#include <SoftwareSerial.h>
SoftwareSerial mySerial(7, 8); //create software Serial
#define rx_Pin 7 //define the software RXD pin
#define tx_Pin 8 //define the software TXD pin
#include <dht11.h>
dht11 DHT; //create a DHT object
#define DHT_Pin 6 //define the humidity & temperature pin
#include <Arduino.h>
#include <Servo.h>
Servo myservo; //create a servo object
#define ESC_Pin 9 //define the ESC output pin
int pulsewidth; //define the pulse width
int temp = 0; //test motor
#define PMArrayLenth 31  //0x42 + 31 bytes = 32 bytes
unsigned char buf[PMArrayLenth]; // air quality data
int PM01Value = 0; //define PM1.0 value
int PM2_5Value = 0; //define PM2.5 value
int PM10Value = 0; //define PM10 value
int HumValue = 0; //define humidity value
int TmprValue = 0; //define temperature value
unsigned int sampleInterval = 2000; //sampling interval is 2s
unsigned int printInterval = 2000; //communication interval is 2s
unsigned int displayInterval = 2000; //LCD display internal is 2s
unsigned long sampleTime, printTime, displayTime;

/* check the data of air quality */
char checkValue(unsigned char *thebuf, char leng) {
  char receiveflag = 0;
  int receiveSum = 0;
  for (int i = 0; i < (leng - 2); i++) {
    receiveSum = receiveSum + thebuf[i];
  }
  receiveSum = receiveSum + 0x42;
  // check the accumation
  if (receiveSum == ((thebuf[leng-2] << 8) + thebuf[leng-1])) {
    receiveSum = 0;
    receiveflag = 1;
  }
  return receiveflag;
}

/*count PM1.0 value of the air detector module */
int transmitPM01(unsigned char *thebuf) {
  int PM01Val;
  PM01Val = ((thebuf[3] << 8) + thebuf[4]);
  return PM01Val;
}

/*count PM2.5 value of the air detector module */
int transmitPM2_5(unsigned char *thebuf) {
  int PM2_5Val;
  PM2_5Val = ((thebuf[5] << 8) + thebuf[6]);
  return PM2_5Val;
}
```

```
/*count PM10 value of the air detector module */
int transmitPM10(unsigned char *thebuf) {
  int PM10Val;
  PM10Val = ((thebuf[7] << 8) + thebuf[8]);
  return PM10Val;
}

/* display the minitoring value */
void LCDdisplay(int pm25, int tmpr, int humi) {
  lcd.setCursor(0, 0); //location of display
  lcd.print("PM2.5=");
  lcd.print(pm25); //display PM2.5 value
  lcd.print("ug/m3");

  lcd.setCursor(0, 1); //location of display
  lcd.print("TMPR=");
  lcd.print(tmpr); //display temprature value
  lcd.print("C");
  lcd.setCursor(9, 1); //location of display
  lcd.print("HUM=");
  lcd.print(humi); // display humidity value
  lcd.print("%");
}

/*define pulse function */
void servopulse(int pin, int val) {
  int myangle;
  // all from 0 - 180 to 500 - 2480
  myangle = map(val, 0, 180, 500, 2480);
  digitalWrite(pin, HIGH); //high
  delayMicroseconds(myangle); //delay
  digitalWrite(pin, LOW); //low
  delay(20 - val / 1000); //delay
}

void setup() {
  lcd.begin(16, 2);  //define rows and columns of LCD
  pinMode(rx_Pin, INPUT); //define RXD input
  pinMode(tx_Pin, OUTPUT); //define TXD output
  Serial.begin(9600); // configure baudrate of serial port
  mySerial.begin(9600); // configure baudrate of software serial
  mySerial.setTimeout(1500);

  pinMode(ESC_Pin, OUTPUT);
  for (int i = 0; i <= 110; i++) {
    servopulse(ESC_Pin, 150); //max speed about 2s
  }
  for (int i = 0; i <= 55; i++) {
    servopulse(ESC_Pin, 20); //min speed about 1s
  }
  for (int i = 0; i <= 150; i++) {
    servopulse(ESC_Pin, 31); //control the motor
  }
  myservo.attach(ESC_Pin);   //define motor i/o
}

```

```
void loop() {
  // read data from sensors
  if (millis() - sampleTime >= sampleInterval) {
    sampleTime = millis();
    int chk = DHT.read(DHT_Pin);  // read humidity & temperature
    HumValue = DHT.humidity;
    TmprValue = DHT.temperature;

    if (mySerial.find(0x42)) { // read air quality data
      mySerial.readBytes(buf, PMArrayLenth);
      if (buf[0] == 0x4d) { // split air quality value
        if (checkValue(buf, PMArrayLenth)) {
          PM01Value = transmitPM01(buf);
          PM2_5Value = transmitPM2_5(buf);
          PM10Value = transmitPM10(buf);
        }
      }
    }
  }

  // display sensors' value using LCD
  if (millis() - displayTime >= displayInterval) {
    displayTime = millis();
    LCDdisplay(PM2_5Value, TmprValue, HumValue);
  }

  // print the information on the serial monitor (Bluetooth).
  if (millis() - printTime >= printInterval) {
    printTime = millis();
    Serial.println(PM01Value);
    Serial.println(PM2_5Value);
    Serial.println(PM10Value);
    Serial.println(TmprValue);
    Serial.println(HumValue);
  }

  // ESC control by the variable speed
  if (PM2_5Value > 70)
    // all from 70 - 150 to 0 - 250
    temp = map(PM2_5Value, 70, 150, 0, 250);
  else
    temp = 0;
  double len = 700; //len = 0.7*1000 ~ 1.0*1000
  len = len + temp;
  myservo.writeMicroseconds(len); //control motor
}
```

Chapter 7
A Fire-Fighting Robot Using Arduino

7.1 Introduction

A robot is defined as a mechanical design that is capable of performing human tasks or behaving in a human-like manner. Building a robot requires expertise and complex programming. It is about building systems and putting together motors, solenoids, and wires, among other important components. There are a number of subsystems that must be designed to fit together into an appropriate package suitable for carrying out a robot's task. The fire-fighting robot competition is an international competition where the challenge is to design and construct a fire-fighting robot that can find its way through an arena that represents a model house, find a lit candle that represents a fire in the house, and extinguish the fire in the shortest time while avoiding any obstacles in the robots path. As the contests web page states, the main purpose of this contest is to "provide an incentive for the robotics community to develop what can be a practical application for a real-world robot." The competition aims to increase awareness of robotic fire-fighting, encourage team-based education, and promotes robotics as a theme for teaching engineering design.

The arena is designed as Fig. 7.1 and the explanations are follows:

(1) The field is set to 3×3 m, and is designed for five rooms. The start position is designed as a black rectangular area.
(2) The door of each room has a black belt. It is convenient for the robot to detect the room location by using the line-tracking sensor.
(3) There is a single lit candle in any one of the five rooms, randomly selected. The robot enters the room, finds the lit candle, and extinguishes it.
(4) The fire-fighting robot returns to the start position safely.

T. Pan and Y. Zhu, *Designing Embedded Systems with Arduino*,
DOI 10.1007/978-981-10-4418-2_7

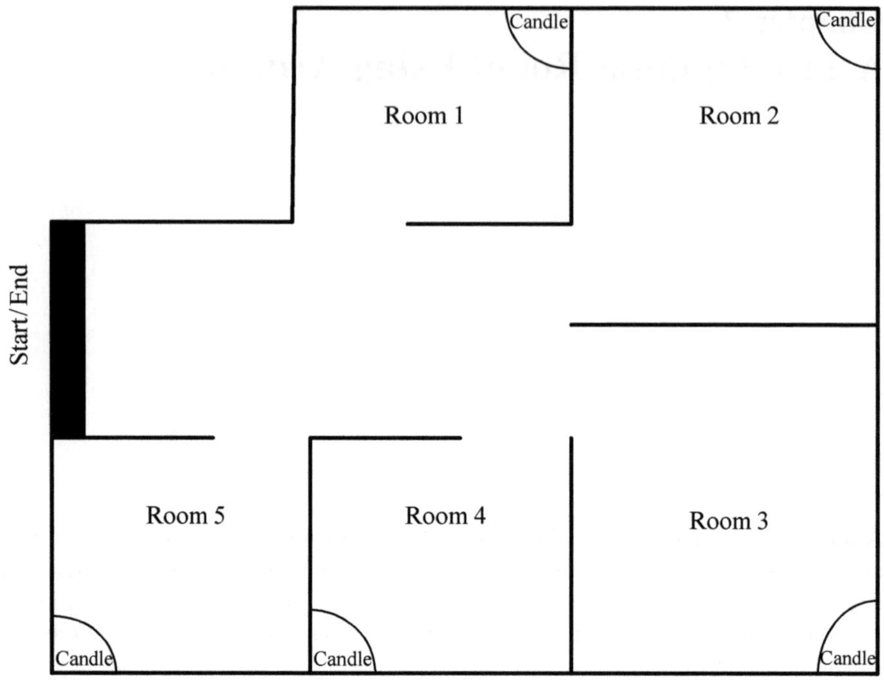

Fig. 7.1 The diagram of arena

7.2 Task Definition

In this competition, the robot needs to finish three tasks: search the fire source, extinguish the lit candle, and return to the start position.

7.2.1 Task 1: Search the Fire Source

In the task of searching the fire source, the intelligent robot should find a lit candle in a room in the shortest time. As the lit candle is placed randomly, the robot should enter each room as fast as possible.

The criteria for the robot to enter the room are that the robot's body should completely cross a pre-drawn black line at the door.

The criteria for successfully searching the lit fire are that the robot prepares to extinguish the fire or already commences fire-fighting operations when it meets the fire source.

7.2.2 Task 2: Extinguishing the Fire

When the robot is close to the location of the fire source, it needs to extinguish the fire source. In this competition, we use a fan to extinguish the fire. It should be noted that the simulated fire source (candles) cannot be knocked down.

7.2.3 Task 3: Returning to the Start Position

After finishing the task of fighting the fire, the robot needs to return back to the start position as soon as possible. Further, when returning, the robot does not need to go into other rooms.

7.3 Robot Design

In order to satisfy the competition requirements and to be able to realize the goal of the competition, which is to detect a lit candle in a house model and extinguish it in the shortest possible time, a fast reliable robot with a body that is able to lift all the hardware and at the same time be light enough to remain fast is required. An easy shape that could navigate through the house while avoiding obstacles is designed first. Here, we simply select a premade four-wheel car as the robot's body (shown in Fig. 7.2), which is purchased from DFrobot (http://www.dfrobot.com.cn/goods-443.html).

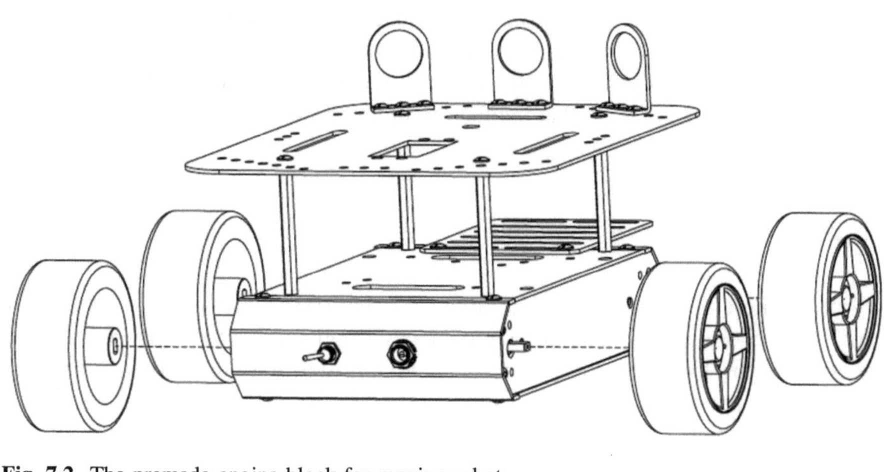

Fig. 7.2 The premade engine block for moving robot

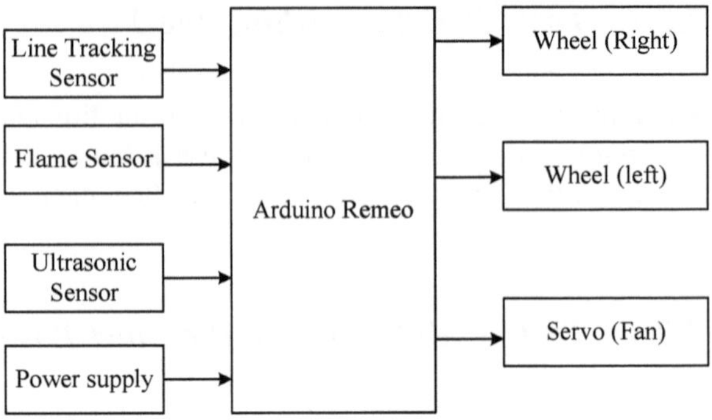

Fig. 7.3 Structure diagram of fire-fighting robot

The system block diagram of the fire-fighting robot is shown in Fig. 7.3.

7.3.1 Sensors

Three kinds of sensors are used, i.e., flame sensor (DFR0076), line-tracking sensor (SEN0017), and ultrasonic sensor (HC-SR04).

The flame sensor is used to detect the fire source. The accuracy of the flame sensor is up to 3 ft. It detects flames or wavelengths of light source within 760–1100 nm (please refer Sect. 3.17 for details).

The line-tracking sensor detects the black belt at the door and informs the robot about how many rooms it has crossed (please refer Sect. 3.5 for details).

The ultrasonic sensor (please refer Sect. 3.6 for details) is used to recognize the walls and avoid obstacles. The intelligent robot reads the distance between the robot and the wall using three ultrasonic sensors at the front, left, and right side of the four-wheel car, and adjusts the distance constantly. Thus, the fire-fighting robot can search for the fire source along the wall and return. Using the ultrasonic sensor at the head of the car body, the fire-fighting robot can measure the distance between the car and the fire source, so that the fan driven by the servo can extinguish the fire.

7.3.2 Extinguishing System

In our method of extinguishing candles, the DFR0332 fan module (shown in Fig. 7.4) is chosen. We chose this fan over other options due to its low cost, small size and weight, low power requirements, and ability to put out candles within and only within a small distance. This fan has three pins to control its operation: VCC, ground, and signal.

Fig. 7.4 The shape of fan module

The driving codes are as follows.

```
#define Fan_Pin 3     //define driver pins

void setup() {
  pinMode(Fan_Pin, OUTPUT);
  Serial.begin(9600);     //Baudrate: 9600
}
void loop() {
  int value;
  for (value = 0; value <= 255; value += 5) {
    analogWrite(Fan_Pin, value);    //PWM
    Serial.println(value);
    delay(30);
  }
}
```

7.3.3 *Motor Drive*

The four wheels are driven by a pair of DC motors that are interfaced to the Arduino Romeo through L-298 dual H-bridges. L298 can drive two DC motors that can be controlled in both the clockwise and anticlockwise direction. It has an output current of 600 mA and peak output current of 1.2 A per channel. The in-build diodes protect the circuit from back EMF at the outputs (please refer Sect. 4.1 for details).

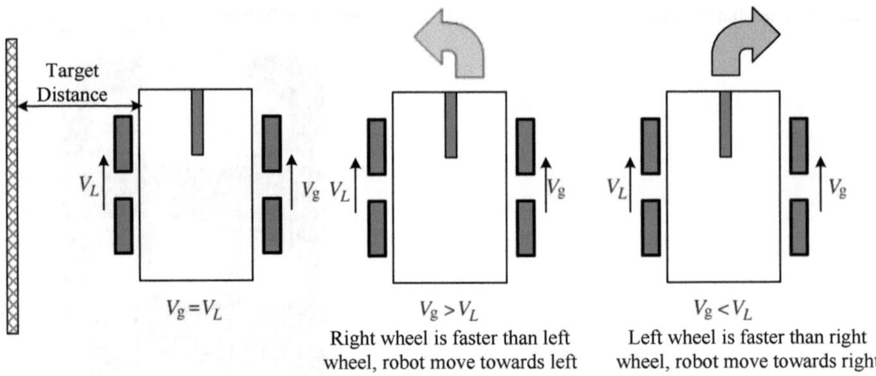

Fig. 7.5 The schematic of wall-following

7.3.4 Algorithms and Behaviors

In this competition, the movement of the robot is ensured such that it maintains a fixed distance from the side walls. Therefore, the speed of the wheel is adjusted such that one wheel is faster than the other. The difference in speed between both wheels helps in varying the movement direction of the robot (Fig. 7.5).

The ultrasonic sensor measures the distance between the robot and wall, and the robot adjusts the distance by using the driving motors. As a result, the robot can walk along the wall. To control the speed of the motor, the pulse width modulation (PWM) technique is used to vary the voltage of the motor. Arduino is responsible for sending a PWM signal to L298 Quadruple Half H Driver, which controls the voltage supplied to each motor (please refer Chap. 4 for details).

7.4 Demonstration

7.4.1 Components

- DFRobot Romeo board and USB cable × 1.
- DFR0076 (flame sensor) × 3.
- SEN0017 (line-tracking sensor) × 3.
- HC-SR04 (ultrasonic sensor) × 3.
- DFR0332 (fan module) × 1.
- Jumper wires × *n*.

7.4.2 Romeo Pinouts Are as Follows

Motors (four outputs required):

4—Right Motor Direction
5—Right Motor Speed
6—Left Motor Speed
7—Left Motor Direction.

Fan (one output required):

3—Fan Pin A.

Ultrasonic sensors (9 output required):

A3—Ultrasonic Left Trig
A4—Ultrasonic Left Echo
8—Ultrasonic Front Trig
9—Ultrasonic Front Echo
0—Ultrasonic Right Trig
1—Ultrasonic Right Echo.

Line-tracking sensors (three inputs required):

11—Line Sensor 1
12—Line Sensor 2
13—Line Sensor 3.

Flame Sensors (three inputs required):

A0—Flame Sensor Left
A1—Flame Sensor Font
A2—Flame Sensor Right.

7.4.3 Results

When the robot first starts, it runs a calibration routine. This establishes base values for the flame and ultrasonic sensors, so that action can be taken when the sensors provide values significantly above the calibration values. Each individual flame sensor and ultrasonic sensor have their own separate calibration value. For more information, please refer Chap. 3.

To realize the goal, i.e., searching the fire source, extinguishing the candles, and returning to the start position, two algorithms are used in the robot, i.e., wall-following algorithm and fire-fighting algorithm.

The wall-following algorithm is that the fire-fighting robot moves according to the left-hand rule. The robot turns right only if there is no other possibility, whereas it always turns to the left if there is an option. Here, L = UltraLeft() and F = UltraFont();

- If L < 10, the robot is close to the wall, and should turn right;
- If 10 < L < 13, the distance between the wall and robot is suitable, and the robot moves forward;
- If L > 13, the robot is far away from the wall, and should turn left;
- If F < 19, there is an obstacle in front of the robot, and the robot should turn right.

When a flame sensor detects a candle, the robot pauses its search algorithm. If the candle was detected by one of the side sensors, the robot turns in the direction and lets the proximity sensor (front) see the candle. After the robot is centered on the candle and completely stopped, the fan turns on. After four seconds the fan turns off, and the robot checks the front flame sensor to see if it succeeded in extinguishing the candle. If it did not, the robot would readjust and repeat by turning the fan on.

7.4.4 Codes

```
#define E1 5 //M1 Speed Control
#define E2 6 //M2 Speed Control
#define M1 4 //M1 Direction Control
#define M2 7 //M1 Direction Control
#define IR1 11   //line tracking sensor 1
#define IR2 12   //line tracking sensor 2
#define IR3 13   //line tracking sensor 3
#define flame_L A0   // flame sensor in left
#define flame_F A1   // flame sensor in front
#define flame_R A2   // flame sensor in right
#define trig1 A3  //ultrasonic sensor in left
#define echo1 A4
#define trig2 8   //ultrasonic sensor in front
#define echo2 9
#define trig3 0   //ultrasonic sensor in right
#define echo3 1
#define fan 3    // extiguishing fan

```

```
int ldistance,fdistance,rdistance;
int Lspeed = 190; //basic speed for left motor
int Rspeed = 180; //basic speed for right motor
int flag = 0; //tasks flag: Searching room, Judging fire, a
Extinguishing fire, Return
int flag2;
int room = 0; // room number

void setup() {
  pinMode(E1, OUTPUT);
  pinMode(E2, OUTPUT);
  pinMode(M1, OUTPUT);
  pinMode(M2, OUTPUT);
  pinMode(echo1, INPUT);
  pinMode(echo2, INPUT);
  pinMode(echo3, INPUT);
  pinMode(trig1, OUTPUT);
  pinMode(trig2, OUTPUT);
  pinMode(trig3, OUTPUT);
  pinMode(IR1, INPUT);
  pinMode(IR2, INPUT);
  pinMode(IR3, INPUT);
  pinMode(fan, OUTPUT);
}

// -----------------------------------------------
// Functions for motor moving
void stopp(float c) { //Stop
  digitalWrite(E1, LOW);
  digitalWrite(E2, LOW);
  delay(c * 100);  //duration
}
void advance(float c) { //forward
  analogWrite (E1, Rspeed + 10);  //PWM Speed Control
  digitalWrite(M1, HIGH);
  analogWrite (E2, Lspeed + 10);
  digitalWrite(M2, HIGH);
  delay(c * 100);  //duration
}
```

```
void back (float c) { // backward
  analogWrite (E1, Rspeed);  //PWM Speed Control
  digitalWrite(M1, LOW);
  analogWrite (E2, Lspeed);
  digitalWrite(M2, LOW);
  delay(c * 100); //duration
}
void left(float c) { //turn left (one motor working)
  analogWrite (E1, Rspeed);
  digitalWrite(M1, LOW);
  analogWrite (E2, LOW);   // left motor stop
  delay(c * 100); //duration
}
void right(float c) { //turn right (one motor working)
  analogWrite (E1, LOW);  //right motor stop
  digitalWrite(M1, HIGH);
  analogWrite (E2, Lspeed);
  digitalWrite(M2, LOW);
  delay(c * 100); //duration
}
void turnL (float c) { //Turn Left (two motors working)
  analogWrite (E1, Rspeed);  //PWM Speed Control
  digitalWrite(M1, LOW);
  analogWrite (E2, Lspeed);
  digitalWrite(M2, HIGH);
  delay(c * 100); //duration
}
void turnR (float c) { //Turn Right (two motors working)
  analogWrite (E1, Rspeed);  //PWM Speed Control
  digitalWrite(M1, HIGH);
  analogWrite (E2, Lspeed);
  digitalWrite(M2, LOW);
  delay(c * 100); //duration
}
// ---------------------------------------------
// Functions for ultrasonic sonser
float UltraFont() {
  float m;
  digitalWrite(trig1, LOW); //Clears the triger Pin
  delayMicroseconds(2);
  // Sets the trigger Pin on HIGH state for 10 micro seconds
  digitalWrite(trig1, HIGH);
  delayMicroseconds(10);
  digitalWrite(trig1, LOW);
  m = pulseIn(echo1, HIGH);
  m = m * 0.034 / 2; // Calculating the distance
  return m; // return the distance
}
float UltraLeft() {
  float m;
  digitalWrite(trig2, LOW); //Clears the triggre Pin
  delayMicroseconds(2);
  // Sets the trigger Pin on HIGH state for 10 micro seconds
  digitalWrite(trig2, HIGH);
  delayMicroseconds(10);
  digitalWrite(trig2, LOW);
  m = pulseIn(echo2, HIGH);
```

```
  m = m * 0.034 / 2; // Calculating the distance
  return m; // return the distance
}
float UltraRight() {
  float m;
  digitalWrite(trig3, LOW); //Clears the triggre Pin
  delayMicroseconds(2);
  // Sets the trigger Pin on HIGH state for 10 micro seconds
  digitalWrite(trig3, HIGH);
  delayMicroseconds(10);
  digitalWrite(trig3, LOW);
  m = pulseIn(echo3, HIGH);
  m = m * 0.034 / 2; // Calculating the distance
  return m; // return the distance
}
// ---------------------------------------------------
// Function for extinguishing fire
void Fans() {
  int m;
  m = analogRead(flame_L); // if the fire in the left side
  if (m < 800) {
    turnR(0.6);  // adjust the motor to right
    stopp(1);
    m = analogRead(flame_L);
  } while (m > 800);
  stopp(20);

  m = analogRead(flame_R); // if the fire in the right side
  if (m < 800) {
    turnL(0.6); // adjust the motor to left
    stopp(1);
    m = analogRead(flame_R);
  } while (m > 800);
  stopp(20);

  if (analogRead(flame_F) < 800)  { // if the fire in the font
    analogWrite(fan, 250);
    delay(4000);
    analogWrite(fan, 0);
  }
  stopp(10);

  flag2 = 1; // The fire is in this room
  stopp(10);
}

// ---------------------------------------------------
// Function for left-hand law (searching room)
void Left_rule1() {
  do
  { fdistance = UltraFont();
    ldistance = UltraLeft();
    if (fdistance < 21)
      turnR(0.3);
    else if (ldistance < 10)
      right(0);
    else if (ldistance < 13)
```

```
        advance(0);
      else
        left(0);
    } while (digitalRead(IR1) == LOW && digitalRead(IR2) == LOW &&
digitalRead(IR3) == LOW);

    // searching the room and enter in
    if (digitalRead(IR1) == HIGH || digitalRead(IR2) == HIGH ||
digitalRead(IR3) == HIGH)
    {
      back(1);
      stopp(20);
      room++; // record the room
      advance(1.5);   //enter room
    }
    flag = 1;
}
// ---------------------------------------------------
// Function for left-hand law (searching fire)
void Left_rule2() {
    do
    { fdistance = UltraFont();
      ldistance = UltraLeft();
      if (fdistance < 21)
        turnR(0.3);
      else if (ldistance < 10)
        right(0);
      else if (ldistance < 13)
        advance(0);
      else
        left(0);
    } while (digitalRead(IR1) == LOW && digitalRead(IR2) == LOW &&
digitalRead(IR3) == LOW);

    // searching the fire
    if (digitalRead(IR1) == HIGH || digitalRead(IR2) == HIGH ||
digitalRead(IR3) == HIGH)
    { back(1);
      stopp(40);
      if (analogRead(flame_L) < 800 || analogRead(flame_F) < 800 ||
analogRead(flame_R) < 800)
      { Fans();        // it is fire
        advance(1.0);
      }
      else
        advance(2.1);  // no fire
    }
    flag = 2;
}

// ---------------------------------------------------
// Function for left-hand law (leaving room)
void Left_rule3()
{ do
    { fdistance = UltraFont();
      ldistance = UltraLeft();
      if (fdistance < 21)
```

```
      turnR(0.3);
    else if (ldistance < 10)
      right(0);
    else if (ldistance < 13)
      advance(0);
    else
      left(0);
  } while (digitalRead(IR1) == LOW && digitalRead(IR2) == LOW &&
digitalRead(IR3) == LOW);

  if (digitalRead(IR1) == HIGH || digitalRead(IR2) == HIGH ||
digitalRead(IR3) == HIGH)
  { stopp(10);    //prepare to leaving
    advance(1.0);
    if (flag2 == 0)  // if no fire, continue to searching
      flag = 0;
    if (flag2 == 1) // it is fire, go back home
      flag = 3;
  }
}

// -----------------------------------------------
// Function for back (in Room 3 or Room 4)
void Left_rule_Back() {
  if (room == 3)
  { do
    {
      fdistance = UltraFont();
      ldistance = UltraLeft();
      if (fdistance < 19)
        turnR(0.3);
      else if (ldistance < 10)
        right(0);
      else if (ldistance < 12)
        advance(0);
      else
        left(0);
    } while (digitalRead(IR1) == LOW && digitalRead(IR2) == LOW &&
digitalRead(IR3) == LOW);

    stopp(20);
    back(8);
    turnR(6);
    do {
      advance(0);
    } while (digitalRead(IR1) == LOW && digitalRead(IR2) == LOW &&
digitalRead(IR3) == LOW);

    back(1);
    stopp(20);
    delay(10);
    stopp(50);
  }

  if (room == 4)
  { advance(8);
    turnL(7);
```

```
    do {
      advance(0);
    } while (digitalRead(IR1) == LOW && digitalRead(IR2) == LOW &&
digitalRead(IR3) == LOW);
    do {
      advance(0);
    } while (digitalRead(IR1) == LOW && digitalRead(IR2) == LOW &&
digitalRead(IR3) == LOW);
    back(1);
    stopp(20);
    delay(10);
    stopp(50);
  }
}

// -----------------------------------------------
// Function for back (in Room 1 or Room 2)
void Right_rule_Back() {
  if (room == 1)
  { advance(6);
    turnR(5);
    do {
      advance(0);
    } while (digitalRead(IR1) == LOW && digitalRead(IR2) == LOW &&
digitalRead(IR3) == LOW);
    back(1);
    stopp(20);
    delay(10);
    stopp(50);
  }
  if (room == 2)
  { advance(1);
    do
    {
      fdistance = UltraFont();
      rdistance = UltraRight();
      if (fdistance < 19)
        turnL(0.3);
      else if (rdistance < 10)
        left(0);
      else if (rdistance < 13)
        advance(0);
      else
        right(0);
    } while (digitalRead(IR1) == LOW && digitalRead(IR2) == LOW &&
digitalRead(IR3) == LOW);
    stopp(20);
    back(7);
    turnL(7);
    do {
      advance(0);
    } while (digitalRead(IR1) == LOW && digitalRead(IR2) == LOW &&
digitalRead(IR3) == LOW);
    back(1);
    stopp(20);
    delay(10);
    stopp(50);
```

```
  }
}

// -----------------------------------------------
// Main Function
void loop() {
  switch (flag) {
    case 0:
      Left_rule1();
      break;
    case 1:
      Left_rule2();
      break;
    case 2:
      Left_rule3();
      break;
    case 3:
      if (room == 1 || room == 2)
        Right_rule_Back();
      else if (room == 3 || room == 4)
        Left_rule_Back();
      break;
  }
}
```

Chapter 8
Intelligent Lock System Using Arduino

8.1 Introduction

Today, intelligent locks are a hot topic in smart home or smart security systems. Differing from traditional mechanical locks, an intelligent lock always uses non-mechanical identification technologies as the unlocking method, such as fingerprint, Iris recognition, radio frequency card, etc. With the rapid development of mobile Internet, a new method to access and secure private spaces has emerged; people find it more convenient to unlock private spaces using a smartphone as a key. When a specific App is installed in your smartphone, the smartphone becomes a skeleton key immediately. If you want to open a door, you only need to obtain the authority to open it in the App. After obtaining the authority from the administrator, you can open the door without any physical key, but rather your smart phone. It is a desirable alternative.

In this chapter, we demonstrate the design of an intelligent lock based on Arduino, PC or Raspberry Pi, and Android smartphone. This system uses QR code as an unlocking identification, and it can effectively solve the problem of "one key versus one lock" and provide a flexible management method for updating a validation key. The following contents of this chapter focus on the detailed system design, including how to open an electric bolt using Arduino, how to generate a QR code based on Android system, and how to recognize a QR code using PC or Raspberry Pi. After completing this chapter, the user can further develop more convenient and secure intelligent lock systems by themselves.

8.2 System Design

Our demonstration system includes three parts, as shown in Fig. 8.1.

T. Pan and Y. Zhu, *Designing Embedded Systems with Arduino*,
DOI 10.1007/978-981-10-4418-2_8

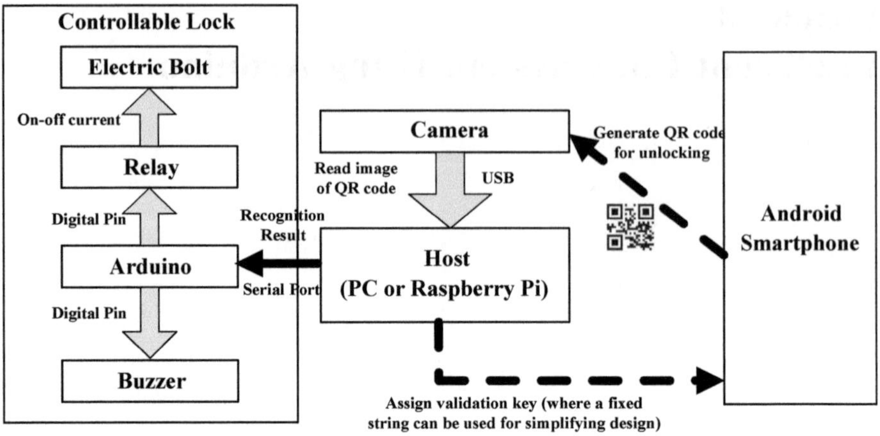

Fig. 8.1 Structure diagram of the indoor air quality monitor

1. A controllable lock: it consists of an electric bolt, Arduino, and relay. Through relay, Arduino can control the electric bolt to open or close it.
2. An application running on Android system: it is used to open the lock as a skeleton key. This APP should have two basic functions: (1) it can obtain a validation key from the host through Wi-Fi or 4G network; and (2) it can generate a QR code for unlocking according to the validation key.
3. A host running on Windows or Linux system: it can be a personal computer or Raspberry Pi. The web camera is connected to the host using a USB cable and Arduino is connected to the host through a serial port. As a server, the host is in charge of generating a dynamic validation key for the Android App, recognizing the QR code from the image captured by the web camera, and then sending an unlock indicator to Arduino after successful validation.

The unlocking methodology of our demonstration system is illustrated in Fig. 8.1. In this system, the electric bolt is controlled by the recognition result of the QR code scanned from a web camera, and the Android App replaces the traditional mechanical key. Therefore, the major processes include five steps.

Step 1: Android App applies authorization (validation key) from host when we click a button on its user interface.
Step 2: Android App generates QR code according to the validation key. Then, a QR code image will display on the screen of your Android phone.
Step 3: Then show this QR code image to a camera connected to the host.
Step 4: The host recognizes the validation key from an image and validates it.
Step 5: The host feedbacks the validation result to Arduino. If the validation key is correct, Arduino will control relay to open the electric bolt. Otherwise, a buzzer connected to Arduino will buzz for warning.

8.2.1 *Key Design of Controllable Lock*

As mentioned above, the controllable lock consists of an electric bolt, Arduino, relay, and buzzer. In this section, we focus on the principle of electric bolt and the usage of relay.

1. Electric Bolt

An electric bolt is an electronic control lock, and the extension or retraction of its lock tongue is determined by the on–off status of input current. When the input current is broken, the lock tongue will retract. According to the number of input wires, the electric bolt can be classified into four categories: two-wire, four-wire, five-wire, and eight-wire. Among these categories, the two-wire electric bolt is used most frequently and we also select it for our demonstration system. Figure 8.2 shows a picture of a two-wire electric bolt. As its name indicates, the two-wire electric bolt has two input wires, neutral wire (black) and live wire (red). The neutral wire should be connected to ground (GND), and the live wire should be connected to the power supply. If any one of these wires is cut off, the lock tongue will retract automatically, and then the door can be opened. Because it can be easily controlled by an input current, it is frequently used on various occasions as a controllable door. Further, it is also suitable for an occasion with high security requirement owing to its characteristic of implicit installation.

2. Relay

A relay is an electrically operated switch, which has a control system (also known as the input circuit) and the controlled system (also known as the output circuit), usually used in automatic control circuit. It is an "automatic switch", using a small electric current to control a larger current. Therefore, the relay has the effects of automatic adjustment, security, conversion circuit in the circuit, and so on.

Fig. 8.2 Photo of electric bolt

Fig. 8.3 Photo of 16A relay module

Relays are used where it is necessary to control a circuit by a low-power signal (with complete electrical isolation between control and controlled circuits), or where several circuits must be controlled by one signal.

In the demonstration project, we select a 16A relay module of Arduino suite, and this module can be used to control solenoid valves, lamps, and other equipment. Obviously, it is suitable for controlling the electric bolt. It can be controlled through the digital I/O port of Arduino (Fig. 8.3).

Specification
Contact Rating (Res. Load): 16A 277VAC/24VDC
Maximum switching voltage: 400VAC(NO)
Max. switching current: 16A
Max. switching power: 4700VA
Operate time (at nomi. Vot.): 10ms max
Release time (at nomi. Vot.): 5ms max
Type: Digital
Single relay board
Digital Interface
Control signal: TTL level.

Relay Module Pinout: There are a total of 7 pins on the relay module board.

(1) Link Arduino Side: Signal, VCC, and GND.
(2) Link Appliance Side (as shown in Fig. 8.4):

- COM (IN): Input positive wire from appliance
- N/A (NC): Not connected
- NC (OUT1): Normally closed, which means that when the relay is off (a digital low “0” is received from Arduino) the device is on
- NO (OUT2): Normally open, which means that when the relay is on (a digital high “1” is received from Arduino) the device is on.

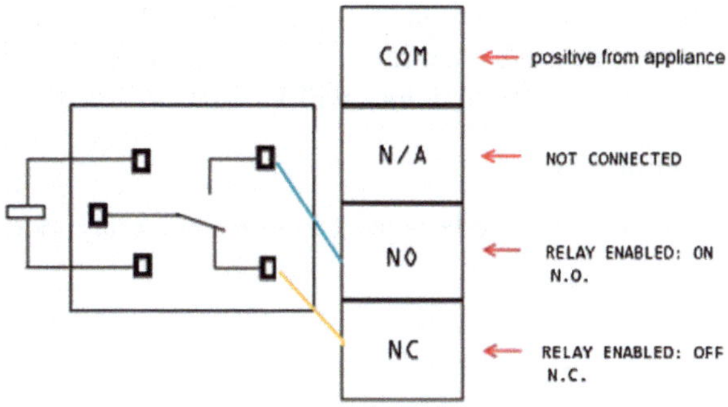

Fig. 8.4 Diagram of connection of relay

WARNING: Please be very careful not to play with live circuits! 120 V or 220 V should not be taken lightly. Make sure the appliance to be tinkered with is unplugged from the mains.

Plugging in Electric Bolt or Lamp

We will use "out1" for connecting the electric bolt, use "out2" to simply reverse the logic, as explained above. Figure 8.5 gives an example of plugging in an appliance such as a lamp, we can replace the lamp to the electric bolt, and the connection relationship is the same.

Step 1: Cut and strip a portion of the positive wire so that you end up with two ends of the wire as shown in Fig. 8.5.
Step 2: The relay should have a positive wire of the device being used connected to "IN" and to "Out 1" as shown in Fig. 8.5, and any digital signal pin on the Arduino end (For example pin 13).

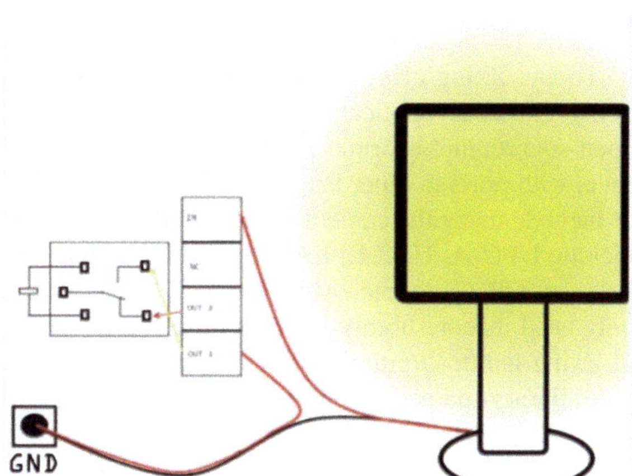

Fig. 8.5 Diagram of plugging in electric bolt or lamp

Step 3: Sending a digital high or a "1" will trigger the relay. Sending a digital low or "0" will disable the relay. If the relay is disabled, the lock tongue of electric bolt will retract; then, the door will be opened.

Sample Code

This sample code is used to test if the relay module works normally.

```
int Relay = 3;
void setup() {
  pinMode(13, OUTPUT);          //Set Pin13 as output
  digitalWrite(13, HIGH);       //Set Pin13 High
  pinMode(Relay, OUTPUT);       //Set Pin3 as output
}
void loop() {
  digitalWrite(Relay, HIGH);    //Turn off relay
  delay(2000);
  digitalWrite(Relay, LOW);     //Turn on relay
  delay(2000);
}
```

8.2.2 Key Design of Android APP

In our demonstration system, the main functions of the Android App include two points. The first point is to obtain the validation key from the host through wireless network. The second point is to generate a QR code for unlocking using the validation key. If we use a fixed string as a validation key, such as "aaaaa", we can simplify the first function. Therefore, we only focus on how to generate a QR code based on Android system in this section.

1. Introduction to Zxing Library

Today, Zxing class library is the most commonly used class library for developing QR code applications on Android system. Zxing ("zebra crossing") is an open-source, multi-format 1D/2D barcode image processing library implemented in Java, with ports to other languages. Zxing can easily realize encoding and decoding of barcode using the camera of smartphone, and the barcode formats supported by it include UPC-A, UPC-E, EAN-8, EAN-13, code 39, code 93, Code 128, Codabar, RSS-14, QR Code, etc. Although the Zxing library has several class libraries, we only need its core library if we just use it to generate QR code. The official website of Zxing is https://github.com/zxing/zxing/releases, and we can also download the core library named "core-3.2.1.jar" from http://central.maven.org/maven2/com/google/zxing/core/3.2.1/core-3.2.1.jar.

【Generate QR code using Zxing core library】

Preliminary 1 Create an Android project, and then add the core-3.2.1.jar to the folder “libs” under the project, Fig. 8.3 shows the operation screenshot.

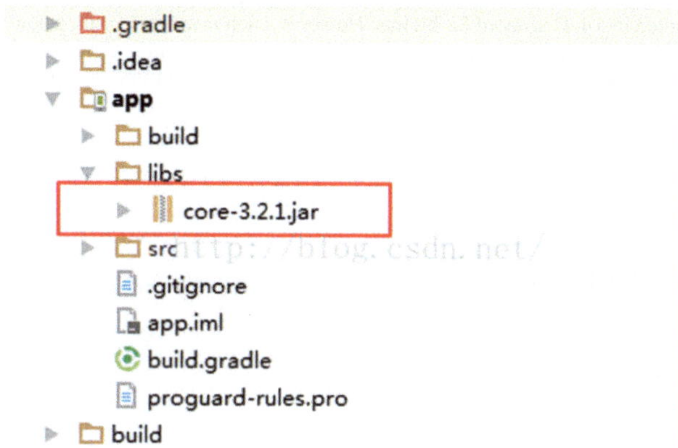

Preliminary 2 After successfully adding the jar package, we can begin to program for generating a QR code. Because QR code is a bitmap image, the target of this program is to generate a special bitmap according to an input string (this string is the information hidden in the QR code). We will use a class named “QRCodeWriter” to realize this function.

Class QRCodeWriter exists within “com.google.zxing.qrcode”, and it has an important method “encode” for encoding barcode. The detailed description of method “encode” is as follows.

Public BitMatrix encode(String contents, BarcodeFormat format, int width, int height, Map<EncodeHintType,?> hints)
throws WriterException

Description: Encode a barcode using input parameters and return a QR Code as a BitMatrix 2D array of greyscale values.

Specified by:
Encode in interface Writer

Parameters:
contents - The contents to encode in the barcode
format - The barcode format to generate
width - The preferred width in pixels
height - The preferred height in pixels
hints - Additional parameters to supply to the encoder

Returns:
BitMatrix representing encoded barcode image

Throws:
WriterException - if contents cannot be encoded legally in a format

This method is used to encode a barcode using input parameters, and its return type is class “BitMatrix”, a 2D array of greyscale values.

com.google.zxing.common

Class BitMatrix

Represents a 2D matrix of bits. In function arguments below, and throughout the common module, x is the column position, and y is the row position. The ordering is always x, y. The origin is at the top-left. Internally the bits are represented in a 1-D array of 32-bit ints. However, each row begins with a new int. This is done intentionally so that we can copy out a row into a BitArray very efficiently.
The ordering of bits is row-major. Within each int, the least significant bits are used first, meaning they represent lower x values. This is compatible with BitArray's implementation.

Constructor and Description
BitMatrix(int dimension)
BitMatrix(int width, int height)

Modifier and Type	Method and Description
void	**clear**() // Clears all bits (sets to false).
BitMatrix	**clone**()
boolean	**equals**(**Object** o)
void	**flip**(int x, int y) // Flips the given bit.
boolean	**get**(int x, int y) // Gets the requested bit, where true means black.
int[]	**getBottomRightOnBit**()
int[]	**getEnclosingRectangle**() This is useful in detecting the enclosing rectangle of a 'pure' barcode.
int	**getHeight**()
BitArray	**getRow**(int y, **BitArray** row) A fast method to retrieve one row of data from the matrix as a BitArray.
int	**getRowSize**()
int[]	**getTopLeftOnBit**() // This is useful in detecting a corner of a 'pure' barcode.
int	**getWidth**()
int	**hashCode**()
static **BitMatrix**	**parse**(**String** stringRepresentation, **String** setString, **String** unsetString)
void	**rotate180**() Modifies this BitMatrix to represent the same but rotated 180 degrees
void	**set**(int x, int y) Sets the given bit to true.
void	**setRegion**(int left, int top, int width, int height) Sets a square region of the bit matrix to true.
void	**setRow**(int y, **BitArray** row)
void	**unset**(int x, int y)
void	**xor**(**BitMatrix** mask)Exclusive-or (XOR): Flip the bit in this BitMatrix if the corresponding mask bit is set.

Example: Based on the usage of “encode” and “BitMatrix”, the following program snippet demonstrates how to generate a QR code according to the input string.

```
Private Bitmap generateBitmap(String content,int width, int height)
{
    QRCodeWriter qrCodeWriter = new QRCodeWriter();
    Map<EncodeHintType, String> hints = new HashMap<>();
    hints.put(EncodeHintType.CHARACTER_SET, "utf-8");
    try {
        BitMatrix encode = qrCodeWriter.encode(content,
BarcodeFormat.QR_CODE, width, height, hints);
        int[] pixels=new int[width*height];
        for (int i=0;i<height;i++) {
            for (int j=0;j<width;j++) {
                if (encode.get(j,i)){
                    pixels[i*width + j] = 0x00000000;
                } else {
                    pixels[i*width + j] = 0xffffffff;
                }
            }
        }
        Return Bitmap.createBitmap(pixels, 0, width, width, height,
Bitmap.Config.RGB_565);
    } catch (WriterException e) {
        e.printStackTrace();
    }
    Return null;
}
```

Explanation: This program snippet defines a function generateBitmap(), and it has three input parameters, where the first parameter indicates the text content hidden in the QR code, and the second and the third parameters denote the width and height of output 2D bitmap, respectively. In this function, we declare an object named “qrCodeWriter”; first, it is an instantiation of class QRCodeWriter. Then, using “qrCodeWriter.encode” method, the input text “content” will be converted to a 2D array. The return result is stored in the variable “encode”, and its type is class BitMatrix. Each element within the 2D array “encode” corresponds to a pixel of output 2D bitmap. We use method “encode.get” to extract the information of each element. If the return value of “encode.get” is true, it means the element has useful data and its corresponding pixel in output bitmap should be black (the RGB representation of black in hexadecimal format is 0x00000000); otherwise, the corresponding pixel in output bitmap should be white (the RGB representation of white in hexadecimal format is 0xffffffff). Next, to create the bitmap using the method “Bitmap.createBitmap”, a 1D array pixel[] is defined to store the color of each pixel, where we adopt two nested for-loops to assign value for pixels[] by going through 2D array "encode". After constructing and filling up array pixels[], a QR code image can be created by the method "createBitmap" (the class Bitmap is realized within android. Graphics). There are six input parameters of the createBitmap method: the first parameter is the colors array to write to the bitmap; the second parameter is the index of the first color to read from pixels[] (a offset value); the third parameter is the number of colors in pixels[] to skip between rows (normally this value will be the same as the width of the bitmap, but it can be larger); the fourth parameter is the number of colors to copy from pixels[] per row; the fifth parameter is the number of rows to write to the bitmap; and the sixth parameter is the color mode of the bitmap (we can set this parameter as RGB_565 as the QR code image only has two colors, black and white).

Call Function generateBitmap(): Now, we add a button and an imageView to the UI of our project; then, we can call the QR code generation function generateBitmap() in the click event of the button. If the id name of imageView is "iv", the sample code snippet is as follows.

```
Public void generate(View view) {
    Bitmap qrBitmap = generateBitmap("http://www.ujs.edu.cn", 400,
400);
    iv.setImageBitmap(qrBitmap);
}
```

8.2.3 Key Design of Host

In above section, we assume that a default string is adopted as a validation key to avoid complex communication programming between the Android client and host. Therefore, we only discuss how to recognize QR codes in this section. If the recognition result is equal to the default string, the host will pass the verification and notify Arduino to disable relay for unlocking the door.

Because the host can be a PC or a Raspberry Pi, it has two available operation systems: Windows and Linux. Based on these two operation systems, many programming languages can realize the function of QR code recognition, such as C#, VB, C++, java, python, etc. Considering that the programming language in the Android side is java, now we continue to use java and the Zxing library to complete the recognition of QR code.

【Read QR code using the Zxing core library】

Preliminary: Within the Zxing library, there is a class named "MultiformatReader", which belongs to "com.google.zxing". By default, the class MultiformatReader attempts to decode all barcode formats that the Zxing library supports. We will use the method decode() of MultiformatReader to read QR code. The detailed description is as follows.

com.google.zxing

Class MultiformatReader

MultiFormatReader is a convenience class and the main entry point into the library for most uses. By default it attempts to decode all barcode formats that the library supports. Optionally, we can provide a hints object to request different behavior, for example only decoding QR codes.

Modifier and Type	Method and Description
Result	**decode**(BinaryBitmap image) This version of decode honors the intent of Reader.decode(BinaryBitmap) in that it passes null as a hint to the decoders.
Result	**decode**(BinaryBitmap image, Map<DecodeHintType,?>hints) Decode an image using the hints provided.

Method Detail

public Result decode(BinaryBitmap image)
throws NotFoundException

Description: This version of decode honors the intent of Reader.decode(BinaryBitmap) in that it passes null as a hint to the decoders.

Specified by: decode in interface Reader

Parameters: image - The pixel data to decode

Returns: The contents of the image

Throws: NotFoundException - Any errors which occurred

public Result decode(BinaryBitmap image, Map<DecodeHintType,?>hints)
throws NotFoundException

Description: Decode an image using the hints provided. Does not honor existing state.

Specified by: decode in interface Reader

Parameters:

- image - The pixel data to decode
- hints - The hints to use, clearing the previous state.

Returns: The contents of the image

Throws: NotFoundException - Any errors which occurred

Example: Based on the usage of method decode(), the following Java program snippet demonstrates how to decode a QR code using Zxing.

```
package com.google.zxing.client.j2se;
import java.awt.image.BufferedImage;
import java.io.File;
import java.io.IOException;
import java.util.Hashtable;
import javax.imageio.ImageIO;
import com.google.zxing.BinaryBitmap;
import com.google.zxing.DecodeHintType;
import com.google.zxing.LuminanceSource;
import com.google.zxing.MultiFormatReader;
import com.google.zxing.NotFoundException;
import com.google.zxing.Result;
import com.google.zxing.common.HybridBinarizer;
public class Decoder {
   public static void main(String[] args) {
      File file = new File("c://qrcodeImage.png");//location of
image file
      BufferedImage bufferedImage = null;
      try {
         bufferedImage = ImageIO.read(file);
      } catch (IOException e) {
         e.printStackTrace();
      }
      LuminanceSource source = new
BufferedImageLuminanceSource(bufferedImage);
      BinaryBitmap bitmap = new BinaryBitmap(new
HybridBinarizer(source)); //convert original image to bitmap
      Hashtable<DecodeHintType, String> hints = new
Hashtable<DecodeHintType, String>();
// DecodeHintType encapsulates a type of hint that a caller may
pass to a barcode reader to help it more quickly or accurately
decode it. It is up to implementations to decide what, if
anything, to do with the information that is supplied.
      hints.put(DecodeHintType.CHARACTER_SET, "UTF-8");
// Specifies what character encoding to use when decoding, where
applicable (type String).
      Result result=null;
      try {
         result=new MultiFormatReader().decode(bitmap, hints);
// Obtain the recognition result
      } catch (NotFoundException e) {
         e.printStackTrace();
      }
      System.out.println(result.toString());
   }
}
```

The above example shows how to read the QR code using Zxing in java, and the return result is string "result". Then, we can compare it with a default validation key using the if statement. When the result matches, the host should notify the successful message to Arduino.

Relay controlling in Arduino:

```
int relay = 3; // connect relay to digital pin 3
int volume=12; // connect buzzer to digital pin 12
char i;
void setup() {
  Serial.begin(9600); // set baud rate of serial port as 9600bps
  pinMode(relay, OUTPUT); // set digital pin 3 as OUTPUT mode
  pinMode(volume,OUTPUT); // set digital pin 12 as OUTPUT mode
}
void loop() {
  if(Serial.available()>0){
    i= Serial.read(); // read data from serial port
    if(i=='1'){ // if received data is equal to "1", where we use
"1" as the message of passing verification
      digitalWrite(relay, HIGH);
      delay(1000);
      digitalWrite(relay,LOW); // control relay to output low
level, then lock tongue of electric bolt will retract
    }
    else if(i=='2'){ // if received data is equal to "2", where we
use "2" as the message of verification failure
      digitalWrite(volume, HIGH); // control buzzer to buzz for
warning
      delay(1000);
      digitalWrite(volume,LOW);
    }
  }
}
```

In this example, we use "1" as a successful message and "2" as a failure message. That means, the host will send "1" to Arduino through serial port if the validation key is correct, and it will send "2" to Arduino through serial port if the validation key is wrong. From the Arduino side, it should initialize serial port in the setup stage, and then use serial.read() method to obtain data from the serial port. If the received data is equal to "1", Arduino should disable relay; otherwise, Arduino should control the buzzer to buzz for warning.

8.3 Photos of Demonstration System

Based on the above description, we realize a demonstration system using Arduino, Raspberry Pi, Android smartphone, and Fischer robot. Figures 8.6, 8.7 and 8.8 show some photos of our demonstration system. Figure 8.6 shows the main hardware components of this system. We can find the camera located on the top-left area. Further, from top to bottom in the middle area, there is Raspberry Pi, electric

Fig. 8.6 Photo of main hardware components

Fig. 8.7 Photo of entire demonstration system

bolt, and Arduino, respectively. Figure 8.7 shows the photo of the entire demonstration system, and we use the mechanical component of the Fischer robot to build a house for demonstration. In this house, each door has a camera (installed above the door) and an electric bolt (installed in the middle of the door). Figure 8.8 shows the screenshots of the Android App that we developed. Figure 8.8a shows the login interface and Fig. 8.8b demonstrates a generated QR code; it can be scanned by a camera when the smartphone is close to the door.

Fig. 8.8 Screenshot of Android application

Fig. 8.9 Recommended solution for intelligent device design

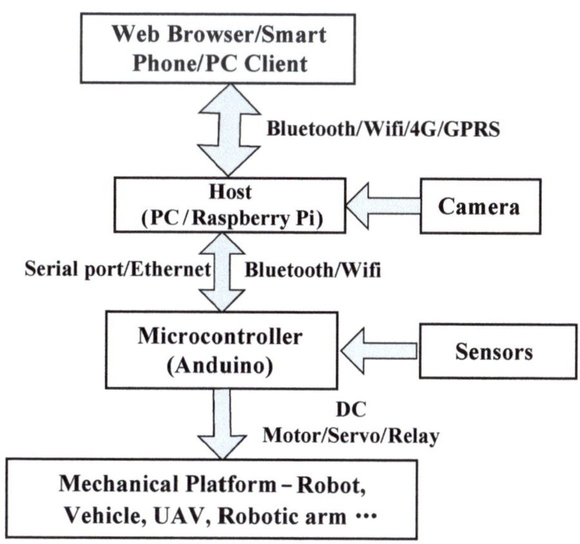

8.4 Conclusion

Based on the demonstration system described in this chapter, we can understand that the most common intelligent device should consist of a mechanical system, microcontroller (such as Arduino), host (such as PC or Raspberry Pi), and smartphone. Microcontroller is important in intelligent device design. Using a microcontroller, the device can receive environment information from sensors, drive DC motor, or servo, control high current equipment by relay. However, this is not enough for designing an intelligent device if we only depend on a microcontroller. Owing to its limited processing capacity, a microcontroller cannot deal with images captured by a camera and cannot meet complex logical processing requirement. Therefore, a host is necessary for the most common intelligent devices. The host has more powerful processing capacities than the microcontroller, and we can program using a high-level language based on its operating system. With the host, the device designed by us can be more intelligent. In the end of this chapter, we provide a solution of intelligent device design, as shown in Fig. 8.9. A user can realize their own designs for any intelligent device according to this solution.

Appendix
Arduino Language Reference

The Arduino language is based on C/C++ and supports all standard C constructs and some C++ features. It links against AVR Libc and allows the use of any of its functions; see its user manual for details.

Structures	
void setup()	The function is called when a sketch starts. Use it to initialize variables, pin modes, start using libraries, etc. The setup function will only run once, after each power up or reset of the Arduino board
void loop()	After creating a **setup**() function, which initializes and sets the initial values, the **loop**() function does precisely what its name suggests, and loops consecutively, allowing your program to change and respond. Use it to actively control the Arduino board
Control structures	
if (condition) { }	Tests whether a certain condition has been reached. Used in conjunction with a comparison operator
if……else	Allows you to do multiple tests
for(initialization; condition; increment)	Creates a loop for repetitive operations

switch (var) {
case label:
    break;
default:
    break;
}
	Allows you to specify different code that should be executed in various conditions

while()	Loops continuously, and infinitely, until the expression inside the parenthesis, () becomes false
do { } while ()	Works in the same manner as the **while** loop, with the exception that the condition is tested at the end of the loop, as does the **do** loop, which will *always* run at least once
break	Exits from a **do**, **for**, or **while** loop, bypassing the normal loop condition

(continued)

T. Pan and Y. Zhu, *Designing Embedded Systems with Arduino*,
DOI 10.1007/978-981-10-4418-2

(continued)

continue	Skips the rest of the current iteration of a loop (**do**, **for**, or **while**). It continues by checking the conditional expression of the loop, and proceeds with any subsequent iterations
return	Terminates a function and returns a value from a function to the calling function
goto	Transfers program flow to a labeled point in the program
Further syntax	
;	Each statement ends in a semicolon
{ }	Curly braces always come in pairs; they are used to define the start and end of functions, loops, and conditional statements
//	Single line comment
/* */	Multi-line comment
#define	Used to give a name to a constant value
#include<avr/pgmspace.h>	Includes outside libraries in your sketch
Arithmetic operators	
=	Assignment operator stores the value to the right of the equal sign in the variable to the left of the equal sign
+	Addition
-	Subtraction
*	Multiplication
/	Division
%	Modulo
Comparison operators	
x == y	x is equal to y
x != y	x is not equal to y
x < y	x is less than y
x > y	x is greater than y
x <= y	x is less than or equal to y
x >= y	x is greater than or equal to y
Boolean operators	
&&	Logical AND. True only if both operands are true
||	Logical OR. True if either operand is true
!	NOT. True if the operand is false
Constants	
HIGH	When reading or writing to a digital pin, there are only two possible values a pin can take (or be set to): **HIGH** and **LOW**
LOW	
true	Logical levels (result of a comparison): false is defined as 0, true is defined as 1 (but more broadly, anything but 0)
false	
INPUT	Digital pins can be used either as INPUT or OUTPUT. Changing a pin from INPUT to OUTPUT with **pinMode**() drastically changes the electrical behavior of the pin.
OUTPUT	

(continued)

(continued)

Data types	
void	Used in function declarations to indicate that the function returns no information
boolean	A boolean holds one of two values, true or false
char	A data type that stores a character value
unsigned char	An unsigned data type that occupies 1 byte of memory. Same as the byte datatype. The unsigned char datatype encodes numbers from 0 to 255
byte	A byte stores an 8-bit unsigned number, from 0 to 255
int	Integers are your primary datatype for number storage, and store a 2 byte value. This yields a range of −32,768 to 32,767
unsigned int	Unsigned ints (unsigned integers) are the same as ints in that they store a 2 byte value. Instead of storing negative numbers however they only store positive values, yielding a useful range of 0 to 65,535
word	A word stores a 16-bit unsigned number, from 0 to 65536. Same as an unsigned int
long	Long variables are extended size variables for number storage, and store 32 bits (4 bytes), from -2,147,483,648 to 2,147,483,647
unsigned long	Unsigned long variables are extended size variables for number storage, and store 32 bits (4 bytes). Unlike standard longs unsigned longs do not store negative numbers, making their range from 0 to 4,294,967,295
float	Datatype for floating-point numbers, a number that has a decimal point
double	Floating-point numbers are often used to approximate analog and continuous values because they have greater resolution than integers. Floats have 6–7 decimal digits of precision
string	Strings are represented as arrays of type char and are null-terminated
arrays	It is often convenient, when working with large amounts of text, such as a project with an LCD display, to setup an array of strings
Conversion	
char()	Converts a value into the char datatype
byte()	Converts a value into the byte datatype
int()	Converts a value into the int datatype
word()	Convert a value into the word datatype or create a word from two bytes
long()	Converts a value into the long datatype
float()	Converts a value into the float datatype

(continued)

(continued)

Utilities	
sizeof	The sizeof operator returns the number of bytes in a variable type, or the number of bytes occupied by an array
Functions: digital I/O	
pinMode(pin, mode)	Configures the specified pin to behave either as an input or an output. pin is the pin number
digitalWrite(pin, value)	Write a **HIGH** or a **LOW** value to a digital pin
digitalRead(pin)	Reads the value from a specified digital pin. The result will be either **HIGH** or **LOW**
Functions: Analog I/O	
analogReference(type)	The default reference voltage is 5 V. This can be changed to a different type and different resolution using this function
analogRead(pin)	Reads the value from the specified analog pin and returns a value between 0 and 1023 to represent a voltage between 0 and 5 V (for default). It takes about 0.0001 s to read an analog pin
analogWrite(pin,value)	Writes an analog value (PWM wave) to a pin. **value** is the duty cycle: between 0 (always off) and 255 (always on). Works on pins 3, 5, 6, 9, 10, and 11
Functions: math	
min(x, y)	Calculates the minimum of two numbers
max(x, y)	Calculates the maximum of two numbers
abs(x)	Computes the absolute value of a number
pow(base, exponent)	Calculates the value of a number raised to a power
sqrt(x)	Calculates the square root of a number
map(value, fromLow, fromHigh, toLow, toHigh)	Re-maps a number from one range to another. That is, a value of **fromLow** would get mapped to **toLow**, a value of **fromHigh** to **toHigh**, values in between to values in between
Functions: trigonometric	
sin()	Calculates the sine of an angle (in radians). The result will be between −1 and 1
cos()	Calculates the cos of an angle (in radians). The result will be between −1 and 1
tan()	Calculates the tangent of an angle (in radians). The result will be between negative infinity and infinity
Functions: random numbers	
randomSeed(seed)	Initializes the pseudo-random number generator, causing it to start at an arbitrary point in its random sequence
random()	The random function generates pseudo-random numbers
Functions: serial communication	
Serial.begin(9600)	Used to begin serial communications, typically at a 9600 baud rate (bits per second)
Serial.print(val,format)	Prints data to the serial port as human-readable ASCII text

(continued)

(continued)

Serial.println(val)	Prints **val** followed by carriage return
Serial.available()	Get the number of bytes (characters) available for reading from the serial port. This is data that's already arrived and stored in the serial receive buffer (which holds 128 bytes)
Serial.read()	Reads incoming serial data
Serial.write()	Writes binary data to the serial port
Serial.end()	Disables serial communication, allowing the **RX** and **TX** pins to be used for general input and output

References

A. Hayes, *Arduino: A Quick-Start Beginner's Guide* (CreateSpace Independent Publishing Platform, 2017)

M. Banzi, M. Shiloh, *Getting Started with Arduino: The Open Source Electronics Prototyping Platform* (Maker Media, Inc., 2014)

J. Blum, *Exploring Arduino: Tools and Techniques for Engineering Wizardry* (Wiley, London, 2013)

B. Francis, *Arduino: The Complete Beginner's Guide* (CreateSpace Independent Publishing Platform, 2016)

K. Chelli, S. Chavhan, Development of wireless sensor node to monitor poultry farm, in *Mobile Communication and Power Engineering*. (Springer, Berlin, 2013), pp. 27–32

A.K. Dennis, *Raspberry Pi Home Automation with Arduino* (Packt Publishing Ltd, 2013)

R. Faludi, *Building Wireless Sensor Networks: With ZigBee, XBee, Arduino, and Processing* (O'Reilly Media, Inc., 2010)

S. Ferdoush, X. Li, Wireless sensor network system design using Raspberry Pi and Arduino for environmental monitoring applications. Procedia Computer Science **34**, 103–110 (2014)

M. Geddes, *Arduino Project Handbook: 25 Practical Projects to Get You Started* (No Starch Press, 2016)

H. Timmis, *Practical Arduino Engineering* (Apress, New York, 2011)

H. Harun, A. Mohd Zin, A study using internet of things concept towards engineering education. Int. J. Adv. Comput. Sci. Technol. **4**(6), 133–136 (2015)

J. Purdum, *Beginning C for Arduino* (2nd edn) (Apress, New York, 2015)

J-D. Warren, J. Adams, H. Molle, *Arduino Robotics* (Apress, New York, 2011)

J. Bayle, *C Programming for Arduino* (Packt Publishing Ltd., 2013)

T. Karvinen, K. Karvinen, *Make: Arduino Bots and Gadgets: Six Embedded Projects with Open Source Hardware and Software* (O'Reilly Media, Inc., 2011)

P. Kumar, P. Kumar, Arduino based wireless intrusion detection using IR sensor and GSM. Int. J. Comput. Sci. Mobile Comput. **2**(5), 417–424 (2013)

A. Kurniawan, *Getting Started with Matlab Simulink and Arduino* (PE Press, 2013)

G. Lockridge, Development of a low-cost Arduino-based sonde for coastal applications. Sensors **16**(4), 1969–2017 (2016)

M. Margolis, *Make an Arduino-Controlled Robot* (O'Reilly Media, Inc., 2012)

M. McRoberts, B. Levy, C. Wootton, *Beginning Arduino* (Apress, New York, 2010)

M. Riley, *Programming Your Home: Automate with Arduino, Android, and Your Computer* (Pragmatic Bookshelf, 2012)

T. Pan and Y. Zhu, *Designing Embedded Systems with Arduino*,
DOI 10.1007/978-981-10-4418-2

Forums

Arduino, Official Arduino forum website (http://www.arduino.org/forums)
Adafruit Industries, Microcontrollers, Adafruit Products, Arduino, and laser cutting (http://forums.adafruit.com)
AVR Freaks, Programming AVR microcontrollers, using GCC, AVR tutorials (www.avrfreaks.net/phorum)
BILDR, Tutorial discussion, hardware and software help (http://forum.bildr.org)
DIY Drones, Autonomous unmanned aerial vehicles powered by Arduino (http://diydrones.com/forum)
MakerBot Industries, Arduino and 3D printing (http://wiki.makerbot.com/forum:start)
Processing, General Processing discussion, project exhibitions, and library development (http://forum.processing.org)
PureData, All things PD—patches, libraries, and hardware (http://puredata.hurleur.com)
RepRap, More 3D printing and other Arduino Arduino-related stuff (http://forums.reprap.org)
SparkFun Electronics, Electronics, SparkFun products, Arduino, PCB design, and project ideas (http://forum.sparkfun.com)

Tutorials

Adafruit Industries, Beginning electronics, sensors, and other Arduino tutorials (www.adafruit.com/tutorials)
Arduino, Official Arduino tutorial website (http://arduino.cc/en/Tutorial/HomePage)
Tutorialspoint, Arduino Tutorial (https://www.tutorialspoint.com/arduino/index.htm)
Jeremy Blum, Video tutorials from Jeremy Blum (http://jeremyblum.com/category/arduino-tutorials)
SparkFun Electronics, Embedded electronics, surface mount soldering, projects, and other tutorials (www.sparkfun.com/tutorials)
Spooky Projects, Class notes and tutorials from Tod E. Kurt (http://todbot.com/blog/spookyarduino)
Tronixstuff, An ever expanding series of Arduino tutorials from John Boxall (http://tronixstuff.wordpress.com/tutorials)
Wiring, Tutorials for the Wiring platform, similar to and mostly compatible with Arduino (http://wiring.org.co/learning/basics)

MIX
Papier aus verantwortungsvollen Quellen
Paper from responsible sources
FSC® C105338

If you have any concerns about our products,
you can contact us on
ProductSafety@springernature.com

In case Publisher is established outside the EU,
the EU authorized representative is:
Springer Nature Customer Service Center GmbH
Europaplatz 3, 69115 Heidelberg, Germany

Printed by Libri Plureos GmbH
in Hamburg, Germany